D1484114

DATE DUE

MAY 0 8 2010	

BRODART, CO.   Cat. No. 23-221-003

1
Springer Series on Biofilms

Series Editor: J. William Costerton

The Biofilm Primer

Volume Author: J. William Costerton

With 67 Figures, 37 in color

 Springer

Series Editor and Volume Author:

Dr. J. William Costerton
Director, Center for Biofilms
School of Dentistry
University of Southern California
925 West 34th Street
Los Angeles, CA 90089
USA

Library of Congress Control Number: 2006939146

ISSN 1863-9607
ISBN 978-3-540-68021-5 Springer Berlin Heidelberg New York
DOI 10.1007/b136878

Springer is a part of Springer Science+Business Media

springer.com

© Springer-Verlag Berlin Heidelberg 2007

Editor: Dr. Christina Eckey, Heidelberg
Desk Editor: Anette Lindqvist, Heidelberg
Cover design: Boekhorst Design BV, The Netherlands
Typesetting and Production: LE-TeX Jelonek, Schmidt & Vöckler GbR, Leipzig

Printed on acid-free paper   149/3100 YL – 5 4 3 2 1 0

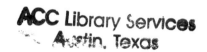

Preface

Most human activities require a framework that may begin with a kindergarten, extend through sports, and culminate in the medieval institution of a university devoted to teaching, scholarly pursuits, and physical and emotional aggrandizement of its members. There is a certain pleasant symmetry in fitting into this framework being seen as a competent scholar, a journeyman athlete, and a member-in-good-standing of a collegial group that teaches bright youngsters and extends the boundaries of human perception. You play the game by its sensible and evolving rules, the endorphins flow, and you pass contented through the "seven stages of man."

I was blessed to have chosen a warm and wonderful wife who would let me disappear to climb a mountain, or write a grant, and then have our wonderful children all excited to "do something interesting" when Daddy returned. Janet Halliwell customized science funding in Canada, my lab at the new and vigorous University of Calgary grew to more than 40 people and multimillion-dollar funding, and Kan Lam managed the whole group so effectively that we drove the biofilm field forward with 38 refereed papers in a single year (1987). The pace was frantic, the team was winning and the atmosphere heady, and we poured over the goal line like a rugby team on steroids. But the rules of the game limited us to detailed incremental papers and tightly referenced reviews, biofilm perceptions jerked forward unevenly with provocative data in fields as diverse as pipelines and veins, and I woke up one morning and realized I was bored.

At the age of 58, and acutely bored with incremental science in the framework of the single investigator lab, I received an exciting invitation to replace the charismatic leader and founder of the Engineering Research Center (ERC) at idyllic Montana State University. The engineers taught me how to bring a field forward by conducting well-designed experiments that allow generalization and by an ingenious iterative process in which you cycle between concepts and applications until they fit. At Montana State the best all-round scientist I will ever know, Ann Camper, let me "poach" the research of good students and postdocs in her lab, so I didn't have a lab of my own but I got to drink coffee with a succession of young geniuses—you know who you are! I was flying again, I consorted with a mobile cluster of "young turks," I brokered ideas among people of the stature of Pete Greenberg and Buddy Ratner, and the biofilm concept that lies at the center of this book began to take shape. It is an engineering concept, with a scientific base, and it is meant to solve practical problems and to provide a coherent rationale for research in the field. Lynn Preston runs the ERC program at the NSF, and she rubs the noses of errant ERC directors in wet newspaper, until they embrace this engineering "systems" approach—bless her.

Hal Slavkin hired me, in the School of Dentistry at the University of Southern California, because he endorses the biofilm concept and wants to see it applied in all fields of dentistry and medicine. This will happen, and the team is being assembled, but the serendipity is awesome because Ken Nealson is here and because USC has made a "cluster hire" of the brightest and best microbial ecologists whose modern techniques are used to analyze the microbial populations of the oceans. So I stand on a peak in Darien, on West 34th Street, from which I can see buildings in which modern microbial ecologists will use molecular techniques to analyze bacterial populations and brilliant engineers will invoke combustion theory to model biofilm growth. From my fourth-floor aerie I can also see buildings in which microbiology students will earn PhDs without ever seeing a real bacterial population under a microscope and in which specimens from biofilm infections will be streaked on agar plates on which they will not grow. All concerned are good people who play by the rules of their academic frameworks, but they operate in isolation. Some of them must be wrong, very wrong, and the consequences are far from trivial. Hence this diatribe. Hence this manifesto. Hence this blueprint for a new framework and this primitive map for a way forward for microbiology.

October 2006   J. William Costerton

Contents

Introduction

The origins of the sciences of microbiology and virology are sharply different from those of other biological sciences. While intrepid explorers dissected animals and studied their behaviors in exotic locations, and English vicars described hedgerow plants in loving detail through their gentle seasons, microbiology emerged from the fetid fever hospitals of Europe in the mid-1880s. In these grim times, when millions were dying of plague and children were suffocating with diphtheria, the objective was not to describe bacteria as biological entities but to control their depredations on the human race. The mindset and the methods of the early heroes of microbiology were distillation of data and reduction to a useful conclusion, and they thought of themselves more as detectives (de Kruif 1926) than as cloistered academics contemplating the structure and habits of viruses and bacteria.

The continuing strength of microbiology and virology and mycology has been and still is in the protection of man, and his domestic plants and animals, from diseases caused by specialized pathogens. For more than a century we have trained hundreds of thousands of medical and veterinary microbiologists, and large numbers of plant pathologists, and this small army has virtually eradicated the diseases whose causative agents they have so assiduously detected and controlled. These microbe hunters were schooled in Koch's postulates (Koch 1884), the first of which demands the isolation of the pathogen in pure monospecies culture (Grimes 2006), and arcane art forms emerged in which practitioners vied with each other to grow specific pathogens in various complex media. Transport media were developed for the recovery of such pathogens as *Legionella pneumophila*, egg-based media were developed for the growth of *Mycobacterium tuberculosis*, and microbiological gatherings came to resemble recipe exchanges. This relentless focus on the recovery and growth of specific pathogens was successful in that vaccines and antibiotics have been developed for the control of virtually every bacterial or viral scourge, and the stated objectives of the early microbiologists have been largely achieved.

The recovery and culture methods that served the disease detectives so well have been much less successful in the study of the structure and behavior of viruses, bacteria, and fungi in the communities in which they actually live. Because bacteria are not visible to the unaided eye, and because light microscopy presented us with mind-numbing complexity, we have trolled through complex bacterial populations and have grown what we recovered in the same cultures used in medical microbiology. In its infancy the field of microbial ecology benefited from this reductionist approach, in that the metabolic machinery of nitrogen fixation could be studied in bacteria re-covered from ecosystems in which this process had been shown to be both operative and important. We studied cellulose digestion by a bacterial species recovered from the bovine rumen, but we found that we could not extrapo-late back to the functional organ in the animal, because this organism was part of a complex community of which we only studied one or two members. The metabolic machinery of cellulose digestion was operative in the cultured organisms, and the active enzymes were the same as those that digest cel-lulose in the rumen, but the metabolic partnerships that control rates and feedback loops in the real system were missing. Marine microbiologists con-cluded that less than 1% of the different bacteria they distinguished on the basis of morphology actually grew in any type of culture, and most of the species groups detected by modern DGGE techniques fail to grow in any type of medium. A junior student at the Center for Biofilm Engineering probably said it most succinctly when she said that recovery and culture is like running a rake through soil and bushes and trees along a trail, shaking the rake above some potting soil, and basing your study on the plants that grow up in the greenhouse at 37 °C.

This book, and the whole series of biofilm books that will be published by Springer, is based on our understanding of the structure and behavior of bacterial communities that is drawn from the direct examination of these communities. We have, in essence, used new microscopic and molecular tech-niques to walk along the path and peer intently at the soil and the plants, and to study the whole complex integrated community, not just the seeds and propagules.

1   Direct Observations

In the traditional microbiological recovery and culture techniques, the assumption is made that each living bacterium in the sample gives rise to a colony, following placement on the surface of agar containing suitable nutrients, and incubation under suitable conditions. This assumption breaks down if the medium or conditions are not permissive for growth, if the cells are aggregated or if several are attached to the same particle, and if any cells are not in a physiological state that permits their rapid growth in the water film on the agar surface. The development of culture systems has usually been driven by our urgent need to grow a particular human pathogen, for purposes of diagnosis and etiological studies, and the system developed by the CDC to grow cells of *Legionella pneumophila* provides an excellent example. When elderly gentlemen sickened and died in that ill-fated hotel in Philadelphia, every effort was made to develop transport media and culture media that would grow this elusive pathogen, and success crowned these labors, but we still cannot grow most of the bacteria in air-conditioning systems. Quite simply, we develop media and culture systems for specific pathogens, as they impinge on our lives, but no one pretends that we can culture all or even most of the bacteria in any given ecosystem. For these reasons, we have developed media and methods to grow most human animal and plant pathogens that cause diseases in which they clearly predominate, but we lack the media and methods to grow more than 1% of the organisms that cause multispecies diseases or simply occupy natural ecosystems. In spite of their narrow focus, these traditional methods have the advantage of yielding continuing cultures of organisms that can be speciated on the basis of their metabolic properties, and whose properties (e.g., antibiotic sensitivity) can be determined in subsequent tests.

Direct observations of microbial biofilms have recently been facilitated by the application of confocal scanning laser microscopy (CSLM), by the development of optically favorable flow cells, and by the proliferation of specific probes to determine species identity and viability. Direct observations of bacterial populations have always constituted the gold standard of bacterial enumeration in natural ecosystems, especially when the cells were stained with acridine orange, but the CSLM now allows us to count bacteria on opaque surfaces. Our ability to visualize bacterial cells on opaque surfaces such as plastics and tissues provides solid and unequivocal data on bacterial numbers, because the observation is direct, but it also provides information on the mode of growth of the organisms. Bacteria may simply adhere to surfaces as individual cells or they may grow in matrix-enclosed biofilms, in which their Brownian motion is constrained and they are separated by distances ranging from 3 to 10 μm. Phase contrast light microscopy can be equally useful in the determination of the numbers and the mode of growth of bacteria if fluid from a single- or mixed-species system is simply passed into a modern flow cell with an optically correct coverslip as one of its structural components. The usefulness of these numerical and spatial data can now be enhanced by the use of antibodies or 16 S-directed oligonucleotide probes to identify cells of a particular species, and by the use of a live/dead probe that determines the membrane integrity of each individual cell. We can now state unequivocally that direct observation techniques yield accurate data on bacterial cell numbers, mode of growth, species composition, and viability in both planktonic and surface-associated microbial populations.

While modern direct microscopy techniques are clearly well honed and ready to replace culture techniques, in the study of the etiology of disease, the new molecular methods that microbial ecologists use in population analyses of natural ecosystems are equally poised for adoption. These molecular techniques share an advantage with culture techniques in that they examine bacterial populations within large volumes and yield data on the relative prevalence of species in whole ecosystems. While polymerase chain reaction (PCR) is not notably quantitative, the denaturing gradient gel electrophoresis (DGGE) technique is more sensitive and more quantitative, and it yields "bands" that correspond to the species that are present in the whole sample (Amann et al. 1995). The DGGE technique is now being widely applied, in medical and dental fields as well as in ecology, and it is being refined by the production of clone libraries (Burr et al. 2006) and by the replacement of simple gels by high-pressure liquid chromatography (HPLC) (Liu et al. 1998). A useful link can now be made between the molecular techniques and direct microscopy, in that DGGE and related methods can yield information on the 16 S rRNA sequences of the species present, so that 16 S rRNA probes can be constructed for fluorescence in situ hybridization (FISH) analysis using direct microscopy. Now that we can map a bacterial population in situ in infected tissues and gather accurate data on the number, species identity, viability, and mode of growth of all of the organisms present there seems to be little value in extrapolating from cultures of the species that happened to grow when the system was sampled.

We sometimes discount direct macroscopic examinations of surfaces, when we are accustomed to high-tech microscopy, but the simple observation that cobble surfaces are covered with clear slime actually alerted us to the preponderance of biofilms in alpine streams. The slime could be recovered by scraping with a penknife, our fingers told us that it was slippery while our noses told us that anaerobes seemed to be absent, and simple observation with a dirt-encrusted field microscope in direct sunlight introduced us to our first natural biofilm! Simple logic encourages us to favor direct observation over extrapolation, but recent studies that document the failure of recovery-and-culture methods tip the balance even more clearly in favor of the new methods of direct observation and molecular analysis. In a recent study of human vaginal microbiology (Veeh et al. 2003) and of "aspectic loosening" of the acetabular cups used in orthopedic surgery (see details in Sect. 4.3), it became apparent that bacteria living in biofilms on healthy or diseased tissues simply fail to grow when they are placed on the surfaces of agar plates. While this failure of biofilm cells to grow on plates is important, our primary contention is that all culture methods are complicated by factors that result in "counts" that are lower than the number of cells actually present, and that direct observation by suitable microscopic methods is the real "gold standard" of quantitative microbiology. My few desultory attempts to explain "most probable numbers" to engineers, who put man on the moon using very real numbers, have met with more confusion than censure, but it is probably high time that we abandon this arcane practice and embrace direct observation.

1.1
The Predominance of Biofilms in Natural and Engineered Ecosystems

Biofilms predominated in the first recorded direct observations of bacteria, when Antonie van Leuvenhoek examined the "scuff" from his teeth, and many pioneers of microbial ecology watched biofilms develop as they placed seawater in glass containers. In fact, ZoBell (1943) noted a "bottle effect" in that colony counts of fresh seawater declined steadily as planktonic (floating) bacteria adhered to glass surfaces and were lost to the bulk fluid. Civil engineers interested in wastewater treatment realized that most of the bacteria that removed organic molecules from sewage lived in sessile populations on surfaces, and they produced elegant models that predicted the efficiency of both biofilms and flocs in nutrient removal. But these isolated observations were not collated and coordinated until we declared the general hypothesis of the predominance of biofilms in natural ecosystems (Fig. 1), using a more rudimentary cartoon, in Scientific American in 1978 (Costerton et al. 1978).

Gordon McFeters and Gill Geesey took advantage of their outstanding physical condition to gallop tens of miles into the alpine zones of the Absorka and Bugaboo mountains, where they plated and cultured water from icy streams crashing down boulder fields (Fig. 2a). These cultures yielded only ±10 bacterial cells per milliliter, but it soon became obvious that rocks

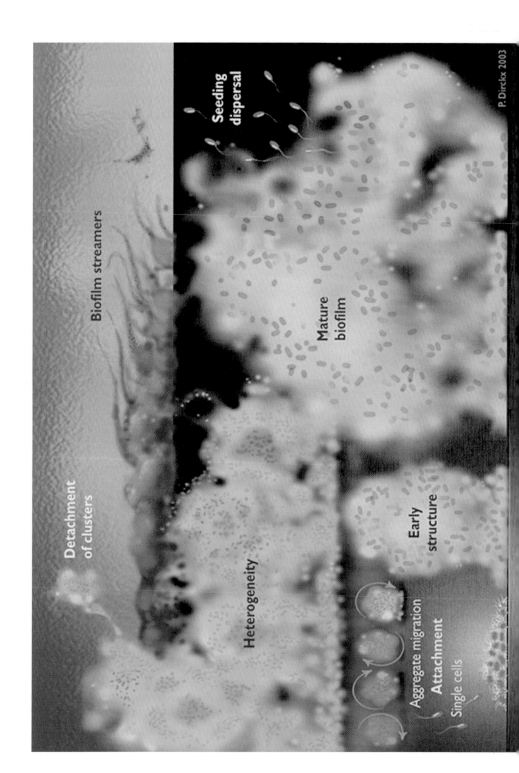

Detachment of clusters

Biofilm streamers

Seeding dispersal

Heterogeneity

Mature biofilm

Early structure

Aggregate migration
Attachment
Single cells

P. Dirckx 2003

◄ **Fig. 1** Comprehensive conceptual drawing showing (*front*) attachment of planktonic cells and sequential stages of biofilm formation, including seeding and detachment. The capability of migration is illustrated (*left*), as is the tendency to form mixed and integrated microcolonies (*middle*) for optimum metabolic cooperation and efficiency. The kelp bedlike configuration of biofilms found in natural aquatic ecosystems (*back*) is also illustrated, as is the tendency of these communities to detach large fragments under shear stress

in the streams were covered with slippery biofilms, and direct examination of these clear slime layers showed the presence of millions of bacterial cells (Fig. 2b) encased in transparent matrices (Geesey et al. 1977). As so often happens in biology, a general truth was revealed by the fortuitous examination of a simple system in which nutrients were severely limited and in which a single species (*Pseudomonas aeruginosa*) formed biofilms on all available surfaces and released a few planktonic cells that were rapidly removed by high flow rates. When we examined a wide variety of rivers and streams, from pristine oilsand rivers (Wyndham and Costerton 1981) to abattoir effluents, this preponderance (> 99.99%) of biofilm cells was sustained in all of these ecosystems (Costerton and Lappin-Scott 1995), and these sessile communities were shown to be proportionately active in nutrient cycling. Biofilms have since been found to constitute the predominant mode of growth of bacteria in streams and lakes in virtually all parts of the world and in the nutrient-rich parts of the ocean, and these sessile populations have been found to be both viable and metabolically active (Lappin-Scott and Costerton 1995; Hall-Stoodley et al. 2004).

Once the tendency of bacteria to form biofilms had been reported, and the appearances of biofilm matrices in light and electron microscopy described (Jass et al. 2003), ecologists reported the presence of biofilms in virtually every natural environment, from tropical leaves to desert boulders. We were inspired to search for biofilms in engineered water systems, with the objective of understanding and controlling processes like corrosion and fouling, because of the enormous cost associated with these problems to the oil-recovery and water-distribution industries. The gradual decay in efficiency of heat exchangers was linked to biofilm formation on the water side of shell and tube units, the removal of these adherent slime layers returned the exchangers to full efficiency, and several companies now ply the biofilm removal trade in industrial water systems. Pipeline engineers had noted that the physical scraping (pigging) was more effective than the use of biocides in the control of microbially influenced corrosion (MIC) in seawater pipelines. The mechanism of MIC was examined, and we found that biofilms on metal surfaces contain areas of differential metal binding capacity and different electrical potentials (Nielsen et al. 1993), and that simple corrosion cell theory can explain how cathodes and anodes within these sessile communities (Fig. 3) can drive MIC at high rates (Lee et al. 1995). Because biofilms mature and begin the MIC process in a matter of weeks, pipeline companies now scrape

Fig. 2 *Top*: alpine stream under Marmolata Spire in the Bugaboo Mountains of southern British Columbia. *Bottom*: TEM of a section through the microbial biofilm that developed on a methacrylate surface immersed in this stream for 30 min. Note the Gram-negative bacterial cells in an ecosystem that grew only *P. aeruginosa* on culture, the extensive matrix composed of exopolysaccharide (EPS) fibers, and the electron-dense clay platelets trapped by the biofilm

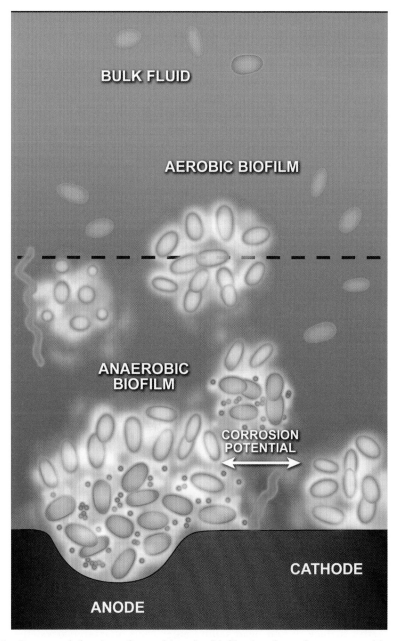

Fig. 3 Conceptual drawing of a multispecies biofilm in whose deeper anaerobic zone a metabolically integrated consortium has developed into an anode, with respect to a neighboring microcolony whose metabolic activities and metal-binding activities have combined to make it relatively cathodic. A corrosion potential has developed between the consortium and the microcolony, in a "classic" corrosion cell, and metal loss occurs at the anode

their lines at regular intervals with pairs of "pigs", with biocide in the intervening fluid, and much less pipe is lost to microbial corrosion. Biofilms also predominate in soils, and the outsides of the same pipes are protected from MIC by the systematic imposition of cathodic protection currents. As we examine more and more ecosystems, from the aerial surfaces of leaves to the ghastly chaos of rumen contents, we always note the predominance of biofilms. We can conclude that the bacteria that live in the biosphere, between the Earth's molten core and outer space, grow almost exclusively in matrix-enclosed communities and that new strategies are urgently needed to study them and to integrate them with the many biological systems currently studied by molecular analysis and direct observation.

Microbial ecologists have embraced the biofilm hypothesis, which states that these sessile communities predominate in the natural and industrial ecosystems of the biosphere, but other bacterial strategies clearly operate in the areas beneath this nutrient-rich crust. Direct observations of the vast nutrient deserts of the deep oceans and the deep subsurface have shown that bacteria adopt a radically different survival strategy in these regions. Dick Morita and his colleagues recovered water from deep oceans and found that it contained very few bacterial cells that could be resolved by ordinary light microscopy, but that the addition of simple nutrients produced direct and culture counts of $\pm 1 \times 10^5$ cells/ml in as few as 20 min (Novitsky and Morita 1976). Further examination produced the fascinating "starvation survival strategy" hypothesis (Fig. 4), which has now been fleshed out and canonized by Staffan Kjelleberg's group (Kjelleberg 1993), in which it is established that starvation triggers the production of very small ($\pm 0.3 \, \mu$m) dormant ultramicrobacteria (UMB). These UMB represent a bacterial mode of growth that is antithetical to the biofilm mode of growth in that the cells are naked, nonadherent, and almost completely metabolically dormant (Fig. 4, top and middle) but capable of resuscitation to form normal vegetative cells (Fig. 4, bottom). UMB have now been found, in approximately equal numbers ($\pm 1 \times 10^5$ cells/ml), in groundwater from as deep as 5000 ft (1500 m) below the Earth's surface, and in the abyssal areas of the oceans. Bacteria can thus be seen to have adapted to Earth's biological realities by adopting the starvation survival strategy in the nutrient-deprived regions of the deep oceans and the deep subsurface and by adopting the biofilm strategy in the nutrient sufficient biosphere. The consequence of this remarkable plasticity of the bacteria is that they exist as a vast metabolically dormant genomic reservoir in the nutrient-poor regions immediately underlying the relatively thin layer at the Earth's surface. When dead sailors enter their Spartan ecosystems, they leap into action and, when currents and deep springs carry them to the surface where nutrients are available, they vie with each other and with existing populations for space and reproductive success.

When rare episodes like the injection of carbon tetrachloride into the subsurface, or the sinking of the Titanic, introduce organic nutrients into the

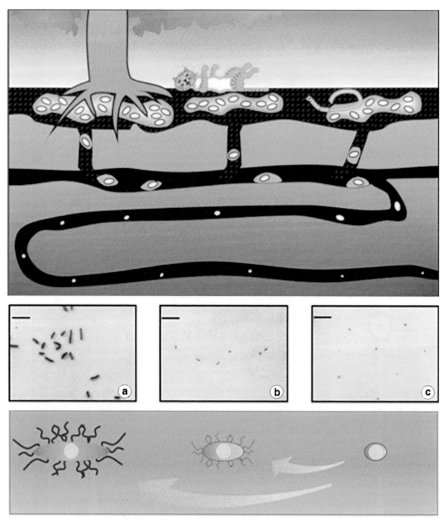

Fig. 4 *Top*: conceptual drawing of biofilm-forming vegetative cells in nutrient-rich upper horizons of soil, which give rise to large numbers of very small starved UMB as planktonic cells are carried down into the nutrient-poor deeper regions. *Middle*: light micrographs of marine vibrio being transformed from vegetative cells (**a**) to much smaller rods (**b**) and to spherical UMB only 0.3 μm in diameter (**c**) by starvation over a 6-week period. From Novitsky and Morita (1976). *Bottom*: cartoon showing resuscitation of UMB to form full-sized biofilm-forming vegetative cells

domain of the UMB, these tiny cells return to their normal vegetative size and resume their tendency to form biofilms (Fig. 4, bottom). We have taken advantage of this starvation-induced shrinkage and nutrient-induced recovery of bacteria to develop a commercial technology for the manipulation of water movement in the subsurface (Fig. 5, top). We select strains of subsurface

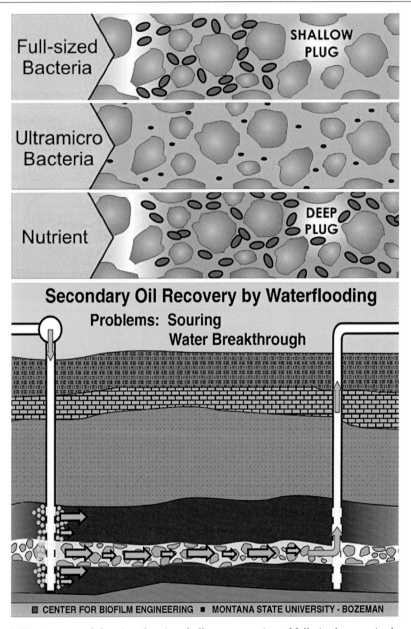

Fig. 5 *Top*: conceptual drawing showing shallow penetration of full-sized vegetative bacterial cells into a porous medium, while UMB can travel (literally) miles through any porous medium > 50 mD in permeability. UMB can be returned to their full size and their full biofilm-forming capability by the addition of nutrients. *Bottom*: this biobarrier technology can be used to plug high-permeability "stringers" that carry injected water past oil deposits, in secondary oil recovery, and the tendency of bacterial biofilms to produce H_2S (*yellow dots*) by the reduction of SO_4 can be controlled by nitrite injection

bacteria, avoiding any tendency to sulfide production or iron deposition, and we grow vegetative cells of the selected strains to very high density in large reactors. The cells are recovered by centrifugation and resuspended in ionically supported distilled water, so that starvation produces very large volumes of suspended UMB that can be transported as stable concentrates. The UMB are injected into the subsurface, where water flow causes problems of pollutant dispersal from point sources, or where the failure of secondary oil recovery is attributed to high permeability "stringers" that carry the injected water past oil reservoirs (Fig. 5, bottom). The UMB are carried as far as 1 km, through any subsurface formation > 50 mD in permeability, and then nutrients are injected by the same route and pumping is suspended to allow the UMB time to return to the full-sized vegetative state (Cusack et al. 1992) and begin biofilm formation. These biofilm "biobarriers" are currently in commercial use for pollutant containment (Dutta et al. 2005), and this technology offers compelling hope that pollutants can be contained and oil can be recovered from established fields that have been abandoned because they were "watered out" (Fig. 5, bottom) (Cusack et al. 1990).

1.2
The Architecture of Biofilms

When microbial biofilms were first visualized, by light microscopy, individual cells could only be resolved in relatively thin sessile communities, and thick biofilms were difficult to visualize with phase contrast optics, especially when they contained crystalline inclusions. Where individual cells could be resolved, it was clear that they were embedded in a translucent matrix that filled the 3- to 6-μm spaces between the cells (Fig. 6) and limited their Brownian movement. Transmission electron microscopy (TEM) of biofilms showed bacterial cells whose structures resembled those of the planktonic cells, but the exopolysaccharide matrices were severely affected by dehydration and could only be resolved if they were stained with electron-dense ruthenium red (Fig. 7). Scanning electron microscopy (SEM) is bedeviled by even more dehydration artifacts than TEM, and attempts to image biofilms were complicated by eutectic bridges that form between cells when their intervening exopolysaccharides are condensed by dehydration (Fig. 8). These bridges appear to connect the cells in biofilms, and they are almost always misinterpreted as intercellular pili. In short, we knew that bacteria lived predominantly in matrix-enclosed biofilms in all nutrient-sufficient ecosystems, but light microscopy was too primitive to reveal the structural details of these ubiquitous and very successful communities, and electron microscopy was fraught with potentially crippling artifacts.

The structural moment of truth came, 15 years after biofilms were seen to predominate in these ecosystems, when we applied confocal scanning

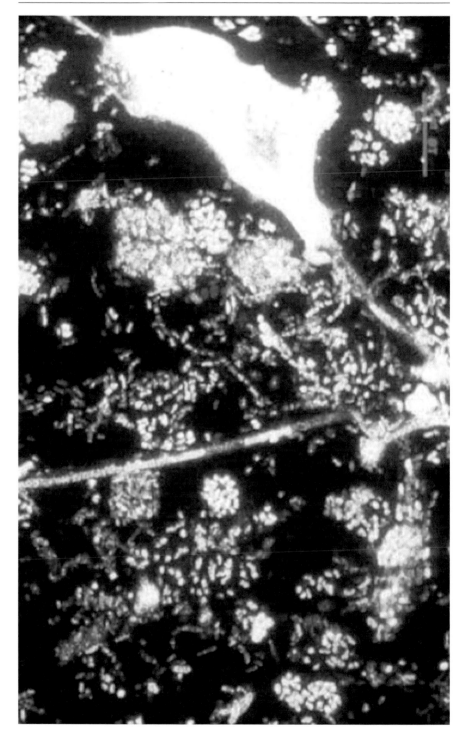

◄ **Fig. 6** Light micrograph of a glass surface immersed in Bow River for 18 h. Note the development of a microbial biofilm consisting of linear trichomes, single bacterial cells, and matrix-enclosed microcolonies within which the sessile cells are separated by several microns. The amoebae seen in this micrograph moved along a trichome, engulfing both single cells and slime-enclosed microcolonies, and the microcolonies were extruded (in a "polished" from) at the trailing end of the protozoan cell

laser microscopy (CSLM) to the study of biofilm architecture. CSLM had been in common use, in most biological sciences, thanks to its ability to produce optical "sections" deep within complex eukaryotic cells, and these sections had often been recombined to produce "maps" of such complex networks as the microtubular cytoskeleton. The fortuitous location of the first biofilm-dedicated CSLM in Doug Caldwell's lab, in Saskatoon, enabled John Lawrence to produce the first confocal images of biofilms, and delegates to the 7th ISME conference, in Kyoto in 1992, were literally buzzing with excitement at their revelations. Sessile cells could be seen to be embedded in a transparent viscous matrix, but the most significant revelations were that biofilms are composed of microcolonies of these matrix-enclosed cells (Fig. 9) and that the community is intersected by a network of open water channels (www.springer.com/978-3-540-68021-5: Movie 1). The movies that accompany this book can be seen on the Springer Web site (http://www.springer.com/978-3-540-68021-5), and expanded versions of the movies can be seen at http://www.usc.edu/biofilms and www.erc.montana.edu. The microcolonies were seen to take the form of simple towers, or of mushrooms, and the water channels were devoid of cells and appeared to constitute a primitive circulatory system that one could imagine being responsible for delivery of nutrients and removal of wastes (Fig. 1). As the delegates returned home from Kyoto and the CSLM paper was published in the Journal of Bacteriology (Lawrence et al. 1991), it was clear that bacteria had taken a very significant step upwards on the ladder of evolution and that these organisms were capable of forming very complex and highly structured multicellular communities (Stoodley et al. 1999b).

When biofilm researchers were given an image of the biological community that we all study, we all began to "twiddle the dials" of culture conditions (Stoodley et al. 1999a), to vary the structure of the sessile communities that developed, and good-natured exchanges broke out between the "lumpy" camp and the "flat" camp. The upshot was that we usually find that well-fed biofilms are unstructured and flat, while less-favored biofilms are highly structured, and (most importantly) biofilms in several natural environments (Fig. 6) are seen to be composed of tower- and mushroom-shaped microcolonies interspersed between open water channels (Møller et al. 1997). The water channels in biofilms inspired the latent hydrologists among the civil engineers in Zbigniew Lewandowski's group to study the flow patterns in this anastomosing network, and convective flow was identified by

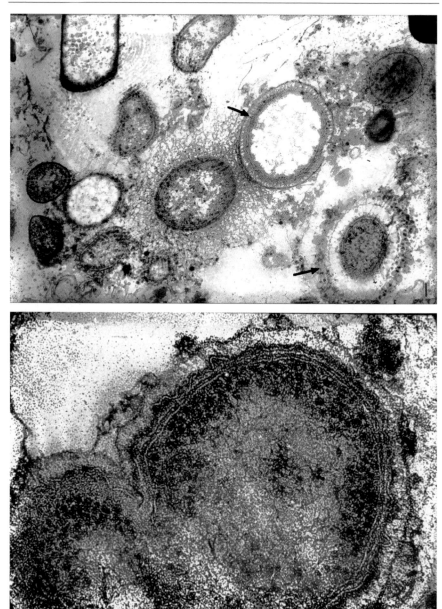

Fig. 7 TEMs of a ruthenium red-stained preparation of bacterial cells living in a competitive functioning ecosystem of the bovine rumen. *Top*: all of these cells are enclosed in very elaborate EPS structures, and the cells in the *12 o'clock* and *4 o'clock* positions (*arrows*) show remarkable concentric reinforcements of their radial EPS fibers. *Bottom*: detail of concentric structure in EPS layer of a different cell. These radial and concentric EPS structures have never been seen in cells in laboratory cultures derived from this very well-studied ecosystem

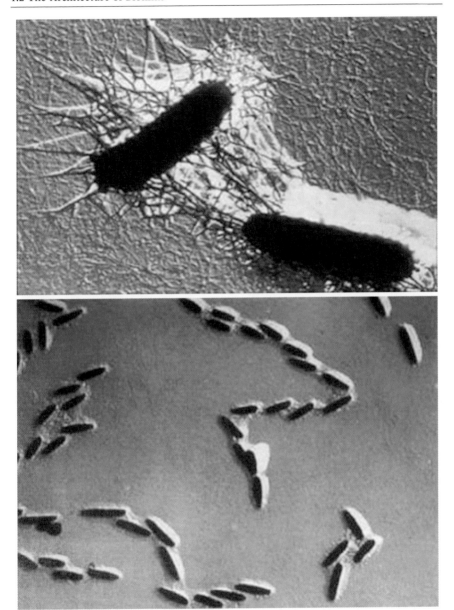

Fig. 8 SEMs of a biofilm that developed on underside of "slick" of synthetic crude oil floated on top of water from Athabasca River in northern Alberta. *Top*: sister cells derived from a bacterial cell that had settled on the oil surface and divided to begin the development of a matrix-enclosed biofilm. The EPS that surrounds these adherent cells is converted to eutectic structures, by the dehydration used in preparation for SEM, and the "fibers" are artifacts that must not be confused with real structures like pili or nanowires. *Bottom*: rapid division of the adherent cells has produced 4-, 8-, and 16-cell clonal aggregates of matrix-enclosed cells on this attractive nutrient surface. (Courtesy Cam Wyndham)

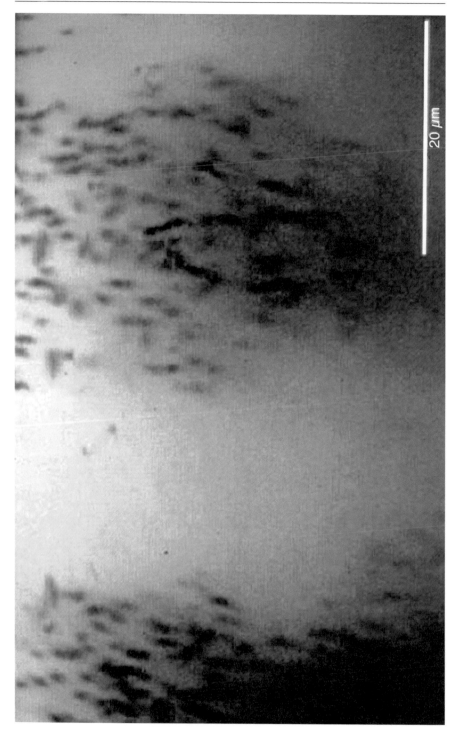

◄ **Fig. 9** Confocal micrograph, in the x–z axis, of a microcolony within a biofilm formed in a flow cell by cells of *P. aeruginosa*. Note the *pale blue* matrix material between the living unfixed bacterial cells and the cell-free water channels that deliver nutrients and remove wastes from this community. (Courtesy Darren Korber)

NMR (Lewandowski et al. 1993) and by direct visualization of the movement of fluorescent particles (Stoodley et al. 1994). The particle studies established the openness of the channels, because particles $> 5\,\mu$m in diameter moved readily through the system (www.springer.com/978-3-540-68021-5: Movie 1), and we later noted that equally large polymorphonuclear leucocytes (PMNs) moved equally readily through water channels (www.springer.com/978-3-540-68021-5: Movie 2). This well-defined architecture of biofilms (Fig. 1) inspired the engineers to test the hypothesis of nutrient delivery via water channels, and microelectrodes were used to map dissolved oxygen concentrations (Lewandowski et al. 1995) in biofilms and showed that this nutrient was indeed delivered to the community via these channels (deBeer et al. 1994).

We get a glimpse of the complexity of cellular distribution within biofilms in the brilliant work of Kjelleberg's group in Australia, in the work of Tolker-Nielsen's group in Denmark, and in their combined work (Webb et al. 2003), and we predict that the distribution of cells within biofilms will eventually be found to be entirely nonrandom. Kjelleberg's group has shown that the marine organism *Serratia liquefaciens* strain MG 1 forms biofilms in which the organism's cells are arranged into vertical stalks that bear rosettes of cells connected to other rosettes by long chains of cells and that each feature of this architectural marvel is controlled by specific genes (Labbate et al. 2004). Tim Tolker-Nielsen's group has shown that one clone of *P. aeruginosa* forms stumplike pedestals on colonized surfaces and that mobile cells of a second clone crawl up the pedestals and form the "caps" of the mushrooms that are such a prominent feature of biofilms formed by this organism (Tolker-Nielsen et al. 2000). At the recent 11th meeting of the International Society for Microbial Ecology (ISME) in Vienna (August 2006) Tim presented evidence (Tolker-Nielsen 2006) that the cells of the second clone may actually form the mushroom caps on templates of DNA produced by the programmed apoptosis of specialized cells in the tops of these pedestals. Even with the stated limitations of in vitro work with single-species biofilms, these studies have special value because they describe the mechanisms and consequences of genetically driven patterns of cell distribution within biofilm microcolonies. If we examine mature biofilms in real ecosystems, we note that the sessile cells are arranged in patterns in which they are separated by standard distances (4 to 10 µm) and that cells may be present in certain parts of towers and mushrooms and completely absent in others. Taken with the observation that sister cells resulting from binary fission are rarely seen together in biofilms, a testable hypothesis emerges in which cells in biofilms are located in genetically determined positions (Fig. 10), much like organelles are

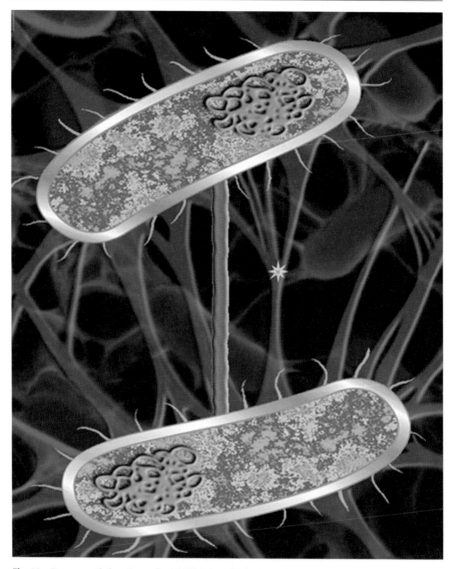

Fig. 10 Conceptual drawing of a biofilm in which the bacterial cells are suspended in an extensive network of pili that connect and position the cells and can contract to bring individual cells together for horizontal gene exchange (see Movie 3). We propose that nanowires also form part of this structural framework, and we venture to suggest that they may be involved in de facto electrical signaling (see spark!) within these structurally integrated communities

located within eukaryotic cells. We are currently engaged in a search for dynamic protein structures that may provide the machinery for specific cell location (www.springer.com/978-3-540-68021-5: Movie 3), and we are encouraged by Satoshi Okabe's recent demonstration (May et al. 2006) that individual cells in *Escherichia coli* biofilms are connected by F pili. All of the cells in these biofilms are connected by multiple pili, and the well-known capability of these structures to contract and apose cells for conjugation can be invoked as a mechanism for other types of localization that may be responsible for the precise positioning of cells within biofilms. Yuri Gorby has found that nanowires are often associated with F pili (Gorby et al. 2006), and the existence of type IV pili in the same communities conjures up an image of a network of at least three kinds of self-assembled protein structures (two of which are contractile) that may position cells within biofilms in the dynamic and controlled manner depicted in Movie 3 (www.springer.com/ 978-3-540-68021-5).

Engineers and mathematical modelers predicted that mushroom-shaped microcolonies would provide optimal diffusion paths for nutrient uptake by sessile bacteria, and our studies of water channels (www.springer.com/978-3-540-68021-5: Movie 1) showed that this system does indeed provide uptake from the bulk fluid and delivery to the community. These comfortable concepts look very convincing on paper (Fig. 1), but we must remember that biofilms are not made of papier mache and that their main structural component is an exopolysaccharide matrix material. Paul Stoodley began to explore the material properties of biofilms by subjecting them to shear forces (Stoodley et al. 2001), and he amazed the biofilm research community with at least one revelation per year, from 1996 until 2002. He showed that individual microcolonies behave like viscoelastic solids and that high shear forces deform them (www.springer.com/978-3-540-68021-5: Movie 4) (Stoodley et al. 1998), cause them to oscillate (Lewandowski and Stoodley 1995), and even cause wave patterns (Stoodley et al. 1999c) to form that traverse the colonized surface and cause large aggregates to detach when the energy of the waves dissipates (www.springer.com/978-3-540-68021-5: Movie 5). All of these conclusions are based on direct observations. The movies showing these behaviors are available at Movies@www.springer.com, and the data have been subjected to rigorous mathematical analysis (Møller et al. 1995). We have attempted to capture the dynamic behavior of biofilms in Fig. 11, which illustrates the oscillating-streamer rolling waves and dynamic detachment processes, but even Peg Dirckx's amazing talents cannot fully capture dynamic processes in two dimensions. So we must conclude that biofilm architecture is essentially ephemeral, in that it is elastic and all of its components respond to stress, and that the architecture that we see at any one point in time is the product of a developmental sequence modified by shear forces. One group of biofilm engineers has even suggested that the towers, mushrooms, and water channels that we see are produced by shear forces and not by directed morphological development. Lest we yield to despair, because the communities we study are so dynamic and protean, we should remember that other multicellular communities (e.g., animals) are equally dynamic

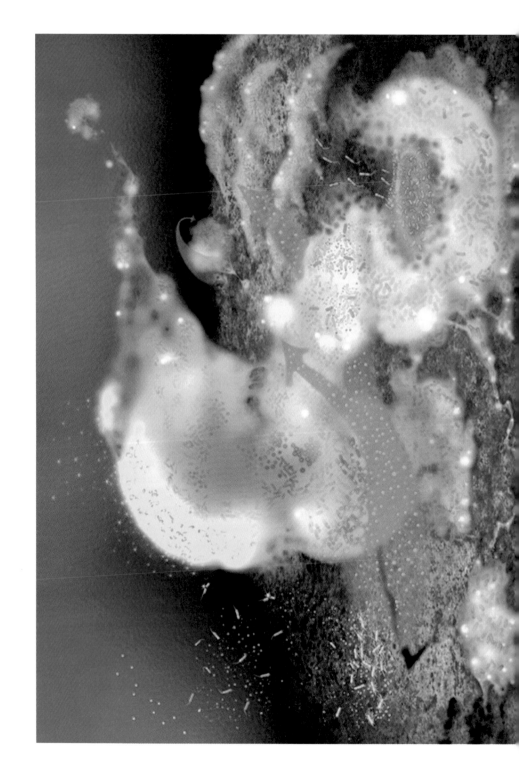

◄ **Fig. 11** In this conceptual drawing Peg Dirckx captures Paul Stoodley's concepts of biofilm dynamics, and the mature biofilm formed by the attachment of planktonic cells (*left*) is capable of moving across the colonized surface in waves (*back right*) and of detaching matrix-enclosed aggregates that may enter the bulk fluid (*top right*) or may roll across the surface (*back center*). The mature biofilm microcolonies may be deformed by shear stress and may also detach planktonic cells (*front right*) that enter the bulk fluid phase. See Movies 5, 9, and 11

and changeable, but simple diagrams (e.g., Figs. 1 and 11) are still useful in considering certain basic processes.

Any consideration of the material properties of biofilms must focus on the matrix material because the cells behave like solid particles and the water in the water channels behaves much like the bulk fluid. So we can conclude that, if the whole biofilm behaves like a viscoelastic solid (Purevdorj et al. 2002), this represents the physical state of the matrix itself. The composition of the matrix is perhaps the most important remaining mystery in biofilm architecture, but we can be sure that the matrices of every biofilm contain certain components. Most matrices stain positively for acid polysaccharides, and those that have been subjected to detailed chemical analysis (Sutherland 1977) have been found to contain polymers of sugar molecules, many of which are uronic acids. Recent studies of natural mixed-species biofilms by Lawrence's group (Lawrence et al. 2003) have shown large "blobs" of exopolysaccharide that don't always enclose bacterial cells per se but do comprise a large part of the volume of these sessile communities. Other direct observations of natural biofilms, by Paul Stoodley and Luanne Hall-Stoodley, have produced images more similar to single-species biofilms grown in vitro, in that most of the sessile cells are actually enclosed by matrix material (Fig. 12). Christoph Schaudinn has used the confocal microscope to examine natural mixed-species biofilms formed on inert "carriers" in the gingival space of periodontitis patients, and his images set a new standard for complexity and artistic beauty (Fig. 13). The truth may lie between these images, and the spatial relations of the cells and matrices of natural biofilms may depend on nutrient conditions in the same way that overall biofilm architecture is influenced by the same factors.

We have always surmised that nucleic acids must be deposited in the matrix when biofilm cells die and lyse, but the revelation that DNA (Whitchurch et al. 2002) comprises a large part of the matrix of some biofilms came as a shock to the biofilm research community. The further revelation (Tolker-Nielsen 2006) that cells at the apices of the mushroom stalks formed by *P. aeruginosa* lyse to release their DNA, which then forms a basis for cap formation by mobile cells of other clones, suggests that DNA may play specific structural roles in biofilm development. One is tempted to speculate that the myxobacterial cells that sacrifice themselves in an equally altruistic manner during fruiting body formation (Kaiser 2004) do so in order to release DNA that plays a pivotal role

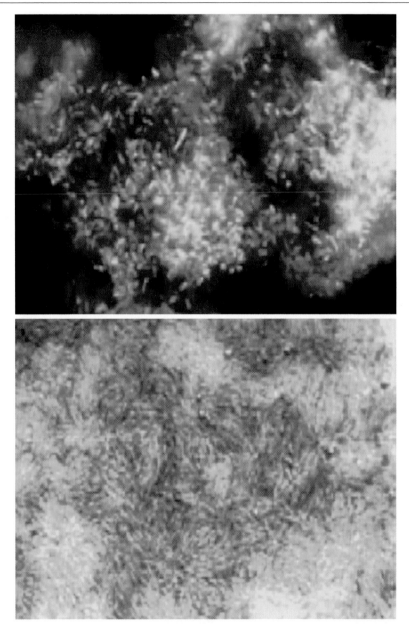

Fig. 12 Confocal micrographs illustrating biofilm formation and simple cell packing. *Top panel*: cells of *P. aeruginosa* in a classic biofilm configuration, in which individual cells are embedded in matrix material, so that all the cells are enclosed and so that cell–cell distances are maintained. *Bottom panel*: simple cell packing by a mutant that lacks the ability to form biofilms, so that there is no matrix material, and the cells are packed together very closely. These shallow layers of packed cells are readily dispersed by surfactants or by shear forces. (Courtesy David Davies)

Fig. 13 Confocal micrograph of biofilm formed on gold foil carrier placed in gingival crevice of patient with controlled periodontitis. Staining with a mixture of confocal probes and fluorophore-tagged lectins shows an arboreal community of linear organisms bearing well-defined bacterial microcolonies, while amorphous EPS material anchors the community. (Courtesy Christoph Schaundinn)

in the development of these bizarre structures. Ulrich Szewzyk's group has very recently published evidence (Bockelmann et al. 2006) that a ramifying network of DNA fibers connects virtually all of the cells of a complex community formed (in vitro) by an organism isolated from "river snow" in South Saskatchewan by John Lawrence's intrepid crew. The physical properties of nucleic acids are not dissimilar to those of polysaccharides, and DNA might be considered a complex polymer of deoxyribose, but the question that most interests us is whether the DNA in the matrix contains information codes or is simply a polysaccharide chain with repeating base units. Tim finds a preponderance of "informational" DNA in a specific area of biofilms, and my original disbelief that bacteria would use "high-investment" DNA for structural purposes is mitigated by my (delayed) realization that the DNA in question has already served its purpose and the producing cells die and release it for "the good of the order". This mental image of the matrix as a tangled mass of various basically polysaccharide polymers would be compatible with the observed viscoelastic properties of the whole community (www.springer.com/978-3-540-68021-5: Movie 4), but it would carry with it the corollaries that the matrix would be permeable to water and would bind large amounts of cations. These corollaries appear to be satisfied by the ATP-FTIR data that indicate that small hydrophilic molecules diffuse through biofilm matrices much as they would through water (Suci et al. 1994) and by electrical data that indicate that large amounts of Mg^{++} and Ca^{++} can be expelled from biofilm matrices by the imposition of a voltage clamp (Stoodley et al. 1997).

We have proposed that, in addition to various polysaccharide polymers and some cellular debris, the biofilm matrix may also contain pili. We are stimulated by two tenuous threads of evidence. We note that sister cells in biofilms separate soon after binary fission and take up positions 3 to 5 μm from each other and that cells are positioned within biofilm microcolonies in patterns that are characteristic of different species. Sometimes the cells are concentrated in the "cap" of the mushrooms and almost totally absent in the "stalk", while other microcolonies of other species display different patterns of cell distribution. These very preliminary observations raise the intriguing possibility that the distribution of cells within biofilm microcolonies is not random but is established and controlled by a network of pili (Fig. 10) that resembles the microtubular and microfibrilar cytoskeleton of eukaryotic cells. If this flight of fancy is true, then it would only be a small extension of this hallucination if the network of pili was thought of as being dynamic, and therefore capable of changing cell distribution in a controlled manner. A further clue indicating that pili may be present and active in the biofilm matrix is that horizontal gene transfer between adjoining biofilm cells occurs at a rate of > 1000 times higher than between planktonic cells suspended in fluids, and conjugation is known to be accomplished by the juxtaposition of cells by contractile pili. Figure 10, and the animations available in Movie 3 (www.springer.com/978-3-540-68021-5), illustrate our suggestion that cell distribution in biofilm microcolonies is controlled by a network of contractile pili, and that one of the functions of these rigid proteinaceous structures is to mediate conjugation.

The recent discovery of very extensive (> 80 μm) "nanowires" that conduct energy from one part of a biofilm to other regions of the community (Gorby et al. 2006) adds functional evidence to the general concept that biofilms are traversed by linear protein structures with myriad functions. The history of microbiology is full of pusillanimous thinking, so I hereby propose that microbial biofilms consist of cells that are connected and positioned by a network of pili (www.springer.com/978-3-540-68021-5: Movie 3) and that the activities of these cells are controlled by cell–cell signaling processes. The cell–cell signals discovered to date are just the "tip of the iceberg", and I predict that we will discover many more signaling systems, and that other types of signals (perhaps electrical impulses – the "spark" in Fig. 10) will be found to be operative. Having watched our concept of bacteria change from individual floating cells to highly structured and metabolically integrated multicellular biofilm communities, I will spend the remainder of my career breathlessly anticipating much more complexity in the microbial world!

1.2.1
Tertiary Structures Formed Within the Matrices of Biofilms

Direct observations of microbial biofilms in natural ecosystems have often shown the presence of regular arrays of walls and partitions, often with a "honeycomb" pattern, but these ordered structures were usually dismissed as being decayed plant materials. Then hexagonal arrays of planar partitions were seen within the biofilms of sulphur-oxidizing bacteria that form very extensive (> 30 mm^2) "veils" on marine sediments (Thar and Kühl 2002) and in microbial biofilms within which calcite is deposited in hypersaline lakes in Bermuda (Dupraz et al. 2004). When similar honeycomb patterns of linear partitions were seen in pure cultures of *Listeria monocytogenes* (Marsh et al. 2003) and of several soil organisms, their microbial origin could not be denied, and we have now seen very extensive arrays of walls and partitions formed by an environmental strain of *P. aeruginosa*. Honeycomb formation appears to be a property that is usually lost when bacterial species are maintained in single-species cultures, but some clones retain this capability during isolation and subsequent cultivation in fluid media. Indeed, it is often instructive to examine old cultures of lab strains that grow as dispersed planktonic cells for the first few days and then form structured networks that fill the entire test tube and can be removed as a single coherent mass.

The bacterial strain that has provided the most unequivocal evidence that prokaryotic organisms can produce very extensive highly organized repeating structures many times their own size is the MH strain of *Staphylococcus epidermidis* isolated from a canine lymphoma by Doug Robinson (Robinson 2005). While most bacterial isolates from these canine tumors produce an organized growth for the first one or two transfers in liquid media and then adopt a planktonic mode of growth, the MH strain retains this capability for an indefinite number of transfers. In liquid cultures the MH strain produces a fine linear network that gradually fills the test tube over a 4-d period and an increasing number of macroscopic white aggregates (Fig. 14a) that form in the liquid and settle in a pellet at the bottom of the tube until a large (± 5 mm^3) mass accumulates by day 4. The basic hexagonal pattern of the network, and the extent of the association of staphylococcal cells with its component fibers, is seen in Fig. 14b,c, and the bona fides of the structure is attested by the demonstration (Fig. 14d) that these structures and their associated cells are clearly seen by light microscopy in unfixed fully hydrated material taken from the test tubes. SEM of the nodes shows that individual spherical cells produce flat plates of an amorphous extracellular material (Fig. 15a,b), as the first stage of the formation of flat tertiary structures, and these flat plates become oriented into extensive walls that form at very regular intervals ca. 8 μm apart (Fig. 15c). When the walls have been formed and are coherent structures, the cells begin to abandon their surfaces and to "build" partitions between the walls at intervals of < 8 μm to form honeycomb structures (Fig. 16a–c) of enormous extent (> 12 000 square μm). When both the walls and the partitions are complete, so that they are coherent and ca. 50 nm thick, the bacterial cells begin to abandon this elaborate system of honeycombs (Fig. 17a,b), leaving the structure seen in Fig. 17c,d. An animated cartoon (www.springer.com/978-3-540-68021-5: Movie 6) in the supplementary material (www.springer.com) illustrates the sequence of structural processes that produces these remarkable tertiary structures in liquid cultures of the MH strain of *S. epidermidis*.

To appreciate the extent to which these observational data have the potential to change the level at which we place bacteria in the hierarchy of living things, we really need to grasp certain facts and look at them without blinking or flinching. First, bacterial cells gather and cooperate to form flat plates of extracellular material (Fig. 15b), and then (somehow) they control the assembly of these small (8- to 10-square-μm) plates into walls that run for hundreds of microns and are separated by a very regular (± 8-μm) space (Fig. 16c). As the walls become well defined and coherent, the bacteria begin to congregate at < 8-μm intervals at locations on either side of the space separating the walls, and then they initiate an annular growth of partitions that link the walls and form a honeycomb pattern with very deep individual elements (Fig. 16c). The ability to form regular tissuelike structures has always been a property reserved for eukaryotic cells, and we have not yet developed an intellectual rubric into which to place the fact that prokaryotic cells can control large-scale activities of this kind. Second, the bacterial cells abandon the honeycombs when the walls and partitions are complete (Fig. 17c,d). This movement of the cells, with the observation that they move to specific loca-

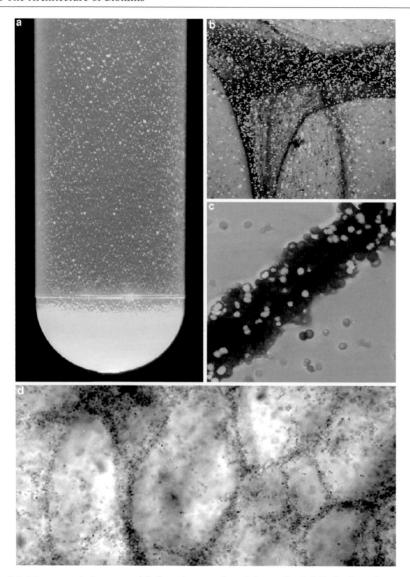

Fig. 14 Macroscopic image and light micrographs of the unfixed fully hydrated network of structures that is formed by the MH strain of *S. epidermidis* growing in liquid culture. **a** Unmagnified image of whole test tube showing formation of a white pellet, and the fine network that extends throughout the culture and contains distinct white nodes whose components are illustrated by SEM in Figs. 15–17. **b** Vital staining of living material shows bacterial cells (*green*) and fibrous components (*blue*) of network on which these cells are suspended. **c** Detail of network fibers realized by overlaying vital stained image with a single confocal "slice" through specimen in which cells are stained green by the use of nucleic acid stain (SYTO 9). **d** Low power light micrograph of unfixed, unstained, living material from the test tube showing association of coccoid cells with very extensive hexagonal structure

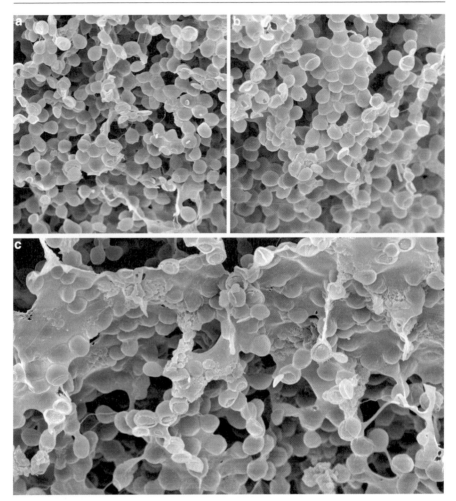

Fig. 15 SEMs of material from nodes of the network formed in a liquid culture of MH strain of *S. epidermidis* showing dispersed single cells (**a**) and gradual assembly of flat plates of amorphous material (**b**) until cells are seen to be associated with very extensive platelike structures (**c**) that may extend for > 50 μm (*arrows*)

tions to initiate partition formation, presumes that they can sense both the completion of structures and spatial locations, and we simply have no perceptual basis to understand these observed activities. We will figure out how this system works, like we determined that gliding motility is mediated by the extension and contraction of pili (Shi and Zusman 1993), and the only thing that is certainly true is that our concept of the complexity of bacterial life in communities will increase exponentially.

The fact that the strain of *S. epidermidis* that retains this ability to form tertiary structures was isolated from a canine lymphoma has stirred speculation,

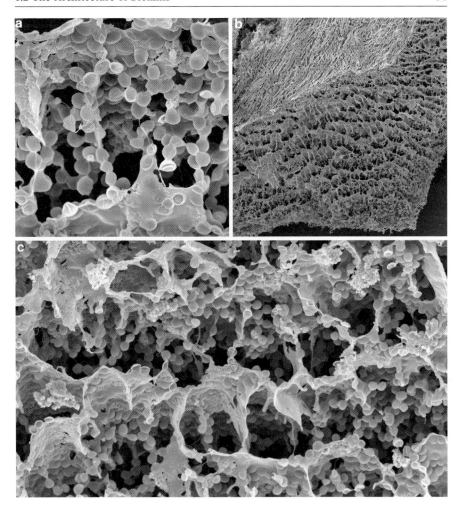

Fig. 16 SEMs of material from the nodes in the network showing the formation of cross "partitions", at ±8-μm intervals, between mature plates. The initiation of this process (**a**) eventually results in the formation of very large honeycomb blocks (**b**) with a detailed honeycomb structure (**c**)

in oncology, that bacteria may play a role in the immobilization of normally mobile lymphocytes to form a solid tumor. *S. epidermidis* is the predominant microbial inhabitant of mammalian skin and has a very long history of association with epidermal tissue, so we will examine the possible role of tertiary structures in maintaining this commensal organism in its peculiar niche. We speculate that the maintenance of the position of nonmotile organisms in a tissue that sloughs relatively rapidly may require some form of anchoring mechanism, and we detect structures (Fig. 18) in human skin that may represent cells of *S. epidermidis* embedded in tertiary structures. Preliminary

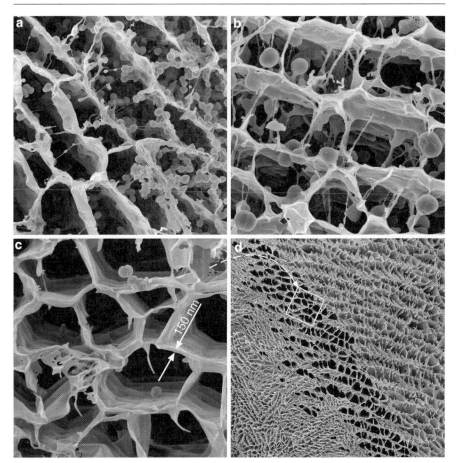

Fig. 17 SEMs of material from more mature nodes showing that the bacterial cells gradually abandon the honeycomb structures (**a**) until the bacterial population is sparse (**b**) or almost nonexistent (**c**), and very extensive areas of honeycomb structure are seen (**d**) to contain only very small numbers of bacterial cells. Figure Fig. 17c shows the detail of the area delineated by the *box* in Fig. 17d, and the macroscopic honeycomb structures built by these bacteria are seen to consist of plates (*arrows*) whose phenomenal level of organization is best seen when the node is bisected at right angles to the deep plane of the honeycomb elements

Fig. 18 Combined image by transmitted light microscopy and epifluorescence microscopy showing mouse tissue that had grown into pores of perforated polyHEMA material that had been implanted in skin for 35 d. The section has been treated with the FISH probe EUB 338, labeled with *Cy3*, and we see a single coccoid bacterial cell (*6 o'clock*) and a microcolony of coccoid cells (*1 o'clock*) that is associated with an extensive area of amporphous material that also reacts with the probe

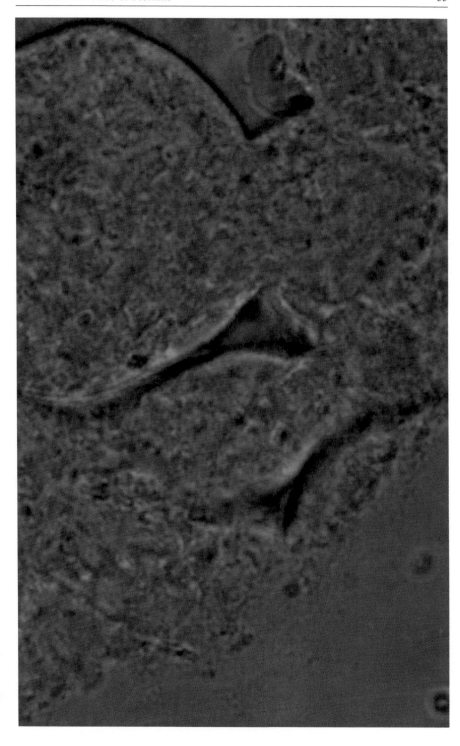

examinations of biofilms in streams have revealed similar tertiary structures, as have direct observations of tissue surfaces in experimental middle-ear infections, and we will now use antibodies to components of bacterial tertiary structures to determine their origins. While we are frankly at a loss to explain why many bacteria make these structures, our working hypothesis is that they function to control variations in environmental conditions in the microniches in which these organisms grow.

1.3
Dynamics of Biofilms

The best and truest observations are often really simple, and a walk beside the Bow River (Calgary, Alberta) can teach us a lot about the bacterial strategy of biofilm formation. The rocks near the sewage plant outfall are coated with a thick grey biofilm that occludes their colors, while the rocks only 50 m downstream bear a thin clear slime with slight green highlights, and the area immediately surrounding fish guts in a backwater is actually gelled by these microbial communities. Biofilms form when planktonic cells encounter organic nutrients, and they develop to thicknesses that reflect the amounts of nutrient available at that site, with the proviso that thick biofilms rapidly become anaerobic and cannot completely oxidize organic molecules to CO_2 and water. So the organic input from a point source will produce a plume of biofilm, on available surfaces, that will extend downstream until it is capable of processing the input from that source. The biofilm plume downstream from a seasonal point source will vary with input, and it can be defined in terms of slime thickness, but the "great leveler" in stream ecology is the scouring of a flood event, and this cataclysm will send the whole biofilm "cartwheeling" downstream as macroscopic flocs. If the biofilm produced by a sewage plant contains pathogens, they will be present in a dormant form but will spring to life in the gut of any unwary drinker of untreated water for miles downstream. If the biofilm contains toxic metals, as is the case at the Grassy Narrows mercury plant in Manitoba (Bodaly et al. 1984), anyone drinking water containing these sinister flocs will suffer their effects for as far as the flocs are transported by the river hydrology.

The biofilm concept is predicated on the fact that planktonic bacterial cells seek to attach to surfaces, by a wide variety of mechanisms (discussed in Sect. 1.3.1), and that these cells may attach irreversibly and undergo the phenotypic changes that initiate biofilm formation (www.springer.com/978-3-540-68021-5: Movie 7). A surprisingly large number of cells that adhere actually leave the surface minutes or seconds after settling, and we really do not understand all of the factors that influence their "decisions". We must realize that adhesion to surfaces in natural mixed-species ecosystems may not follow the patterns seen in studies of biofilm formation by single species in in vitro flow systems, because different species almost certainly interact in their adhesion and biofilm formation strategies. As biofilms develop their fascinating architecture (Fig. 11), and as cells of different species set up their integrated metabolic consortia (Fig. 1, middle), individual cells are caught up in what amounts to an embryological process in which they contribute to overall community function. Even during the "honeymoon" period, during which the nutrient source that elicited the biofilm is maintained, planktonic cells of all component species are freed from their matrix constraints and shed through the water channels into the bulk fluid (www.springer.com/978-3-540-68021-5: Movie 8). These planktonic cells are really a special cohort, shed from successful communities upstream in the same ecosystem and seeking (if bacteria can seek) the same type of nutrient sources on which their home community was founded. In an extreme case, bacterial cells in the bovine rumen become de facto planktonic when the cellulose strands that they digest are fully liquefied, and they float impatiently in the juices of hungry cows waiting for the next bolus of cud containing those identical cellulose strands. By alternating between biofilms and the planktonic mode of growth, in closed systems like the bovine rumen, bacteria can approach levels of efficiency never seen in open natural systems and only duplicated in the best industrial processes.

In open natural ecosystems the issue of predation is added to the issue of nutrient availability, and mature biofilms formed in response to a nutrient input constitute a community that is inherently resistant to antibacterial factors including bacteriophage and amoebae. We have watched as amoebae have grazed within microbial biofilms formed on surfaces in the Bow River (Alberta), and these predatory cells were seen to travel along filamentous bacterial trichomes (Fig. 6) without engulfing any of the cells in these chains. These predatory amoebae ingest and digest free-floating planktonic cells, they can engulf but not process matrix-enclosed biofilm microcolonies, and they cannot even enter large mature biofilms but simply cruise along their borders looking for planktonic stragglers. It is from these heavily defended redoubts that planktonic cells are continuously shed, in all aquatic systems examined to date, and we can think of these undefended skirmishers as scouts searching the downstream ecosystem for freshly exposed surfaces or fortuitous nutrient inputs. Most of these scouts will undoubtedly perish, but those that find a new surface of their liking will set up biofilm communities resembling their home biofilms.

The marine environment exposes bacteria to different stresses and different opportunities. Most of the volume of marine ecosystems has only vanishingly small concentrations of organic nutrients, and bacteria in these abyssal areas simply adopt the starvation survival strategy and reduce their size (Fig. 4) and their metabolic activities to those of ultramicrobacteria. If happenstance provides an input of organic nutrients, specific members of the vast pool of bacterial genomes return to active size and metabolic activity and form biofilms on insoluble organic substrates and on cathodic metals that may serve as sources of free electrons (Nealson 1997; Nealson and Saffarini 1994). In the nutrient-rich surface zones, bacteria form biofilms on inert and nutrient-containing surfaces (e.g., algal fronds), and microbial biofilms on the ghostly shreds of algal fronds form the marine "snow" seen at or near the surface in the Sargasso Sea. The niches within the intertidal zones present a special challenge for which biofilms are very well suited, in that bacteria in these matrix-enclosed microbial communities are highly resistant to drying and to the effects of ultraviolet light. These special properties of biofilms lead to their obvious predominance as macroscopic, and occasionally beautiful, accretions on rocks on craggy headlands where it would seem that the pounding of the surf would preclude any form of life. Perhaps the epitome of marine biofilms are seen in the algal mats that occupy the most salubrious parts of the intertidal zones, and it is useful to remember that marine biofilms shed planktonic cells in the same patterns as stream communities; in addition, all of these sessile communities constantly propagate on new surfaces.

1.3.1
Bacterial Attachment to Surfaces

The true naturalists among us have often taken the best microscopes available to us and taken tours of the real world to see how bacteria behave and grow. These tours were fascinating, but they were recorded by still photography and did not fit the format of scientific publication, so the perceptions they revealed were rarely communicated to the microbiological community. Perhaps every budding microbiologist should spend a week with a Zeiss Photomat, or (better) even a simple confocal microscope, simply watching bacterial cells in the ecosystem whose operations she will plumb for 4 or 5 weary years by extrapolation from cultures. She will then be following in the tracks of creative thinkers in microbial ecology and join Claude ZoBell (ZoBell 1943) in watching the bacteria in mixed natural marine populations assemble in rugbylike scrums on glass surfaces immersed in seawater. She will note that planktonic cells form these twitching phalanxes preferentially on irregularities on the glass and on the surfaces (or even the tracks) of filamentous organisms that have colonized the glass surfaces, and she will further notice that they glide majestically along its smooth valleys. She will join Kevin Marshall and Ralph Mitchell (Marshall et al. 1971) in wondering why individual bacterial cells seem to flirt with an adhesion target by reversible attachment and then "make up their mind" and join the burgeoning microcolony with a "decision" signaled by its cessation of Brownian motion. She will see wonderful things, and she may miss some meals and lose some friends because her fascination with the natural microbial world will skew her schedule and dominate her conversation, but she will never publish these "holiday snaps" and will soon move into the lockstep cloning of genes. Peter Hirsch (University of Kiel) gave us breathtaking views of the natural world with his beautiful photographs of bacterial aggregates in nature (Hirsch and Muller 1986), and Mickey Wagner and Holger Daims led us through biofilms in their favorite cave and stream (Wagner et al. 2003), but these direct observations of bacteria in natural ecosystems are as rare as they are precious. It is surely curious that a biologist, intent on studying a living thing, would not feel compelled to watch that creature as it moves through its normal habitat before dragging it off to the lab and cutting it up.

As biofilm infections have become recognized as the major factor limiting the use of innovative medical devices, the biomaterials community has seized on the notion that some surface characteristic must control bacterial attachment and thus hold the key to infection control. Once the Holy Grail was declared, and a bounty of billions of dollars was attached to it, the race was on and hundreds of scientists from academe and from the "dark satanic mills" of industry set to work. The DLVO theory was invoked (van Loosdrecht et al. 1990), the obvious characteristics of surface roughness and hydrophobicity were identified, and hundreds of papers were published (Fletcher 1987) as the community filled in the complex grids that had bacterial species on one axis and surface characteristics on the other. Then the most myopic of the thundering herd would find a surface property or a bacterial component that affected attachment and would dream that the manipulation of that property or component would control bacterial attachment to biomaterials and even to tissues. At this point avarice would usually triumph, and vice presidents in trench coats would bring us materials that resisted colonization by particular pathogens. We would expose the magic material to freshly isolated bacteria from clinical labs, in the body fluid of choice (often urine), and the lush biofilms that resulted would cause teams that had spend as much as US $2 million of hard corporate cash to disband. What kind of basic intellectual error could cause this continuous lemminglike traffic toward the edge of the same cliff?

Hindsight is always 20/20, but the basic error is that, when we transfer liquid bacterial cultures, we always take a loop from the bulk fluid to inoculate the next (sterile) tube. In so doing we leave behind all of the bacterial cells that have adhered to the surfaces of the test tube, just like ZoBell observed in his "bottle effect" paper (ZoBell 1943), and we gradually select for mutants that are defective in biofilm formation (Fig. 19). Once this process has been repeated hundreds of times, and the resultant lab strain becomes a favorite because it always yields consistent data, so many surface structures have been lost that the cells might actually behave like the inert spherical ball required by DLVO theory in adhesion experiments. When wild strains of E. coli have been transferred ten times in serial culture, they have lost 37.5% of their genes (Fux et al. 2005b), and we must expect that the selective pressure of adhesion to surfaces would have assured the loss of virtually all adhesion factors. So the use of lab-adapted bacterial strains has indicated that changes in some surface property (e.g., hydrophobicity) affect bacterial attachment,

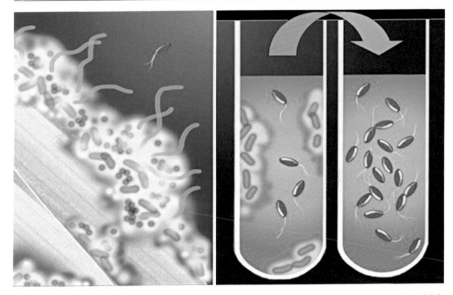

Fig. 19 Conceptual drawing of mixed-species biofilm in the natural ecosystem, in which sessile cells (*pastel colors*) produce occasional planktonic cells (*red*) derived from one clone (*brown*) in a multicellular community. When the single planktonic cell representing one clone of the multispecies community is grown in liquid medium following species selection on an agar plate, subculture techniques leave biofilms on the test tube wall and a small number of subcultures produce strains that have been heavily selected against biofilm formation

and normal corporate secrecy has prevented gossip about research disasters, so I fully expect to be approached by distinguished looking men in trench coats for the remainder of my working life. If the concept of a material that could resist bacterial colonization by virtue of some physical characteristic were correct, we should have found some partial success in the search for this Holy Grail, but three decades of costly research have yielded very little. The take-home lesson is that bacteria are phenotypically plastic creatures that shuffle their genes to adapt to an almost infinite number of different niches and that liquid cultures are an especially nutrient-rich enemy-free niche in which normal transfer methods select against adhesion.

Bacteria growing in real ecosystems are seen to grow predominantly in biofilms, and we note that these sessile communities detach both matrix-enclosed biofilm fragments and free-swimming (or floating) planktonic cells (www.springer.com/978-3-540-68021-5: Movie 8). These biofilm fragments and planktonic cells dominate the bacterial populations of bulk fluids in natural ecosystems and constitute the majority of living entities that would approach an available surface with the potential for attachment. The functional surfaces of a biofilm fragment will be dominated by highly adherent exopolysaccharide fibers, but our direct observations indicate that most biofilm fragments actually lodge in "traps" on surfaces and cannot adhere to surfaces in flowing systems because shear forces operate on them and cause them to roll (Figs. 1 and 11, www.springer.com/978-3-540-68021-5: Movie 9). Individual planktonic cells in natural ecosystems carry pili and exopolysaccharide fibers on their surfaces (Morck et al. 1987), and many also bear flagella whose dynamic actions may cause them to contact all types of available surfaces (Scheuerman et al. 1998), so the receiving surface would "see" many potential bacterial ligands. In our direct examinations of bacterial attachment in mountain streams we noted a pattern in which adhesion was so avid that > 99.99% of cells were attached, and we noted no differences in attachment to leaf tissue, dead wood, or various components of rocks. In real natural ecosystems, and in laboratory experiments using real body fluids and wild bacteria, all planktonic bacteria adhere to all surfaces with remarkable avidity. Roberto Kolter and I have sat (separately) and cried in our wine as bacteria have adhered to the glass wall of our aquaria, and spoiled our home life, and then gone to the office to receive messages that colleagues could not get their favorite lab strains to make biofilms. What is wrong with this picture?

In the approach of an individual planktonic bacterial cell to a surface, the surface would "see" pili and/or flagella made of proteins and 2 to 6 nm in width, and, in wild strains, it would also "see" a mass of exopolysaccharide and lipopolysaccharide fibers enveloping the cell (Figs. 7, 8, and 12). Most wild type cells also have a few vesicles of outer membrane in their matrices, so that some hydrophobic parts of lipopolysaccharides may also be exposed at the surfaces of planktonic cells. Planktonic cells in natural environments survive because they are well protected by a wide variety of surface structures (Fig. 20) (Beveridge 2006), and they present a mosaic of many different components at their surfaces, so that attachment to inert surfaces is surely a multifactorial process. Some bacteria, notably the ETEC organisms that colonize the intestines of young animals, bear surface ligands (e.g., the K99 pilus) that mediate their attachment to cognate receptors on specific tissues (Fig. 21, top), but we must remember that these associations are really only alignments that have no structural strength. When we produced vaccines that raised antibodies against the K99 pilus in *E. coli*, calves were still infected by the K88 strain that uses its capsular material to make equally specific linkages to tissue receptors (Fig. 21, bottom) and produces a strong linkage that resists peristalsis in the gut. Bacteria that can survive in natural ecosystems, teeming with natural chemical and biological hazards, have surface structures of considerable complexity, and they certainly use several mechanisms to accomplish their very avid attachments to inert and biological surfaces. One may ask why the pili that mediate attachment to tissue ligands are so long (2 to 8 μm). And one might observe that structures designed to present bacterial ligands at the effective outer surface of bacterial cells with thick enveloping layers of exopolysaccharide (Fig. 21, bottom) would have to be longer than the matrix was thick.

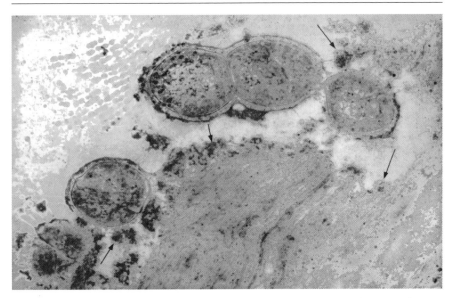

Fig. 20 TEM of bacterial cells in rumen contents. These Gram-positive cells are in the process of digesting plant cellulose, seen as a regular array of parallel fibers on the *right*, and we note the production and deployment of numerous small vesicles (*arrows*) that contain endoglucanases and digest cellulose at some distance from the bacteria

As we take these wild bacteria into the laboratory and subject them to heavy selective pressure toward rapid growth and away from adhesion to surfaces (Fig. 19), they shed their outer structures much like a ripe onion sheds its outer layers and exposes fresh new ones. Figure 7 shows complex outer structures that allow bacteria to survive in the complex rumen ecosystem and have never been characterized because they are rapidly shed in culture. Subculture in liquid media leads to the loss of exopolysaccharide structures, and of some pili and flagella, so that the outer membrane of Gram-negative cells becomes their outermost component, and Gram-positive bacteria may "strip down" to their teichoic acids and peptidoglycans. As an exercise in *reductio ad absurdam*, it is amusing to visualize the most altered and debilitated strain of bacteria that could survive in culture, but not in any natural ecosystem. This strain would need protein synthesis and nucleic acid replication, but mutations in membrane function would be compensated by iso-osmolarity and mutations in community communication, adhesion, protection, and pathogenic mechanisms would not affect its success in culture. Those of us who specialize in cell-envelope structures have often dreamed that some neat component of the outer membrane of Gram-negative bacteria might mediate attachment to target tissues because it is externally located in cultured planktonic cells. Antibodies against these surface ligands have often retarded the adhesion of cultured bacteria to tissues, in carefully constructed animal models in which

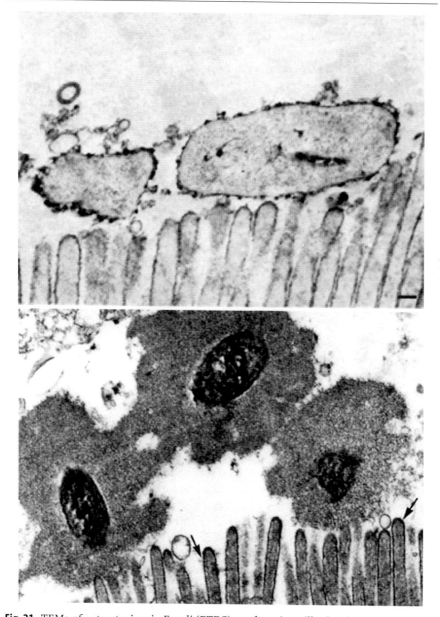

Fig. 21 TEMs of enterotoxigenic *E. coli* (ETEC) on the microvillar borders of experimentally infected calves. *Top*: these pathogens adhere to the microvillar border by means of K 99 pili that can be resolved when they are thickened by reaction with specific antibodies, but their capsules are condensed by the dehydration implicit in TEM preparation and they appear only as electron-dense accretions on the cells and the tissue. *Bottom*: when the K 88 EPS of the bacteria is stabilized by reaction with specific antibodies, the capsules are protected from condensation, and the cells appear in their correct spatial relationships with the colonized tissue

even cultured bacteria can initiate infection, but none of these antiadhesion vaccines has ever been successful in human or veterinary medicine. Studies of the attachment of cultured bacteria to tissue and inert surfaces have raised false hopes in the control of infectious diseases and wasted billions of dollars in the biomaterials industry, but this damage can be contained if we simply resolve to use wild strains in future experiments.

Engineers exercising their penchant for direct observation have examined the behavior of several biofilm-forming strains of bacteria, as they attach to surfaces, and their movies (www.springer.com/978-3-540-68021-5: Movie 10) and models (Hamilton et al. 1995; Heydorn et al. 2000) are readily available at www.springer.com. These are not wild strains, but they have been selected for their ability to form biofilms on surfaces, and they appear to illustrate several general principles of adhesion behavior in bacteria. Andy Rice used complex computer programs to follow the "fate and transport" of planktonic cells of the PAO 1 strain of *P. aeruginosa*, and he showed that only about 50% of the cells that made initial contact with his glass surface remained on that surface for more than a few minutes (Rice et al. 2003). This resembles the "reversible attachment" described in marine organisms by Marshall and Mitchell (Marshall et al. 1971) and suggests that initial attachment may be mediated by London dispersion forces and other readily reversible mechanisms. Cells that remained on the surface moved by the type IV pilus "twitching" mechanism to form microcolonial aggregates (www.springer.com/978-3-540-68021-5: Movie 7) that would eventually begin to produce matrix materials and develop into the microcolonies that constitute the mature biofilm. Andy showed unequivocally that individual cells can and do reverse the attachment process, at any stage of biofilm development, and that even burgeoning young microcolonies shed planktonic cells at a surprisingly high rate. Direct observations of the attachment of planktonic cells of various biofilm-forming species have shown species-specific patterns of cell movement (Korber et al. 1995) following initial adhesion. Cells of some species form circular aggregates, while others form "windrows", and others remain separated and discrete. At this point in time we can see no direct relationship between the patterns that cells form following attachment to surfaces and the architecture of the biofilms that will eventually develop on these colonized surfaces, but we will continue to look for connections.

Bacterial biofilms have the capability of reaching considerable thicknesses because these sessile communities trap nutrients from even the most oligotrophic bulk water, but other factors militate against biofilm accumulation, and most streams and pipes remain functional. Engineers have observed that biofilms formed at low shear can be removed from surfaces by sudden increases in shear force (pressure bumps) but that this susceptibility to increased shear is lost when biofilm remnants that have persisted regrow to reestablish the community. The mechanism of this adaptation to high shear is unknown, but it appears to involve increases in substitution and cross-linking between the polymers that comprise the biofilm matrix. The manner in which bacterial cells sense shear forces and modify their matrices is one of the many biofilm mysteries that await examination. In real stream ecosystems the seasonal variations in biofilms take the form of thin clear accretions during the dark low-nutrient winter months, burgeoning biofilms with algal components as sunlight and nutrients increase, and complete biofilm stripping in episodic floods that move biofilm fragments (including heavy metals and pathogens) downstream in rolling flocs. In more stable ecosystems, like tidal flats and hot springs, microbial mats form the most complex biofilms described to date, and these metabolically integrated communities (Ward et al. 1998) may persist for decades if flow reductions do not cause stagnation and consequent detachment of large portions.

The same forces operate in pathogenic ecosystems; the high shear environment of native heart valves causes the detachment of biofilm fragments (www.springer.com/978-3-540-68021-5: Movie 9) from "vegetations", and these matrix-enclosed aggregates then lodge in capillary beds and form the petticciae that clinicians use to diagnose endocarditis (Olson et al. 1992). As any structural element of the biofilm grows and projects into the bulk fluid, it becomes subject to increased shear forces that operate against the tensile strength of the matrix and its adhesive anchors to the surface, and biofilm migration (www.springer.com/978-3-540-68021-5: Movie 5) and fragmentation (www.springer.com/978-3-540-68021-5: Movie 8) may occur. In the very old iron pipes that still deliver drinking water to parts of our oldest cities, corrosion products and biofilms have narrowed the lumen to less than one fourth its original diameter, but the pumps at the river are powerful and the biofilms near the central channel experience shear forces that still keep the water moving. In this case, and in all cases where thick biofilms have accreted in the lumens of pipes or of medical devices, catastrophic releases of biofilm fragments will result from any ill-advised physical disturbance of these sessile communities. Water engineers have learned to be careful with backhoes, and intensive-care clinicians would be equally well advised to work very gently with wire brushes and even with manipulation and kinking of TPN and ART lines that are heavily fouled with biofilms. Biofilm fragments (www.springer.com/978-3-540-68021-5: Movie 9) constitute a special problem because (unlike detached planktonic cells) they still express the biofilm phenotype, are still enclosed in matrices, and are therefore inherently resistant to conventional antibiotics (Fux et al. 2004) as they move and settle in new locations (Olson et al. 1992).

1.3.2
The Biofilm Phenotype

The first inkling that biofilms were not simply "standard" planktonic bacterial cells embedded in matrices came with the realization that sessile cells are highly resistant to antibacterial agents (Nickel et al. 1985), even though these agents penetrate biofilm matrices with ease. Wright Nichols calculated (Nichols 1991) that small molecules would diffuse readily through the water-filled spaces of predominantly carbohydrate gels, and Jana Jass and Peter Suci (Suci et al. 1994) then used ATR-FTIR to show that specific antibiotics penetrate hundreds of microns of biofilm matrix in seconds. Why, then, were biofilm cells resistant to antibiotics at hundreds of times the concentrations that would kill planktonic cells if biofilms simply consisted of planktonic cells trapped in permeable matrices?

The suspicion that biofilm cells expressed their genes in a pattern that differed from that expressed by planktonic cells, of the same species, was first confirmed by David Davies (Davies and Geesey 1995). David used an elegant reporter construct to show that the *algC* gene was turned on by cells of *P. aeruginosa*, within minutes of their adhesion to a glass surface (Fig. 22), and that the expression of this alginate gene was followed by the visible production of matrix material. The biofilm community has since absorbed and generalized this image of "incoming" bacterial cells "sensing" their proximity to a surface and initiating certain syntheses and behaviors. As we watch cells approaching and colonizing surfaces in monospecies cultures, we should recall that the early giants of microbial ecology watched the same processes as they examined glass slides in seawater containing hundreds of species in natural marine populations. These pioneers established the fact that > 99% of marine bacteria attached themselves to glass surfaces, within an hour of their juxtaposition, and Kevin Marshall and Ralph Mitchell (Marshall et al. 1971) noted that this association went through a reversible phase before it became "irreversible". We have since noted, in studies of pure cultures of *P. aeruginosa*, that as many as 35% of cells that initiate adhesion, postadhesion behaviors, and actual matrix synthesis may leave the surface during the first few hours of biofilm formation.

While swimming cells may impinge on surfaces, floating cells adhere to surfaces using the physical attractions and structures discussed in the section "Attachment" (Sect. 3.1.1), and this initial association with a surface triggers profound and very important changes in the incoming cell. It is a mystery how a bacterial cell "senses" its association with a surface (Prigent-Combaret et al. 1999), but it has been suggested that "hot spots" of signal concentration may develop when the radial diffusion that normally removes these molecules from the producing cells is blocked by the surface. In any event, the association of bacterial cells with surfaces triggers many species-specific behaviors, like the vigorous twitching motility of cells with type IV pili, and other movements that result in the surface-associated cells forming simple clumps or organized "windrows" on the colonized surface. In real ecosystems, in which two or more species may eventually form metabolically cooperative consortia, this movement of cells on surfaces may provide an opportunity for cells of cooperative species to "find" each other and form mixed aggregates before their respective matrices are produced. While the study of surface association and cell mi-

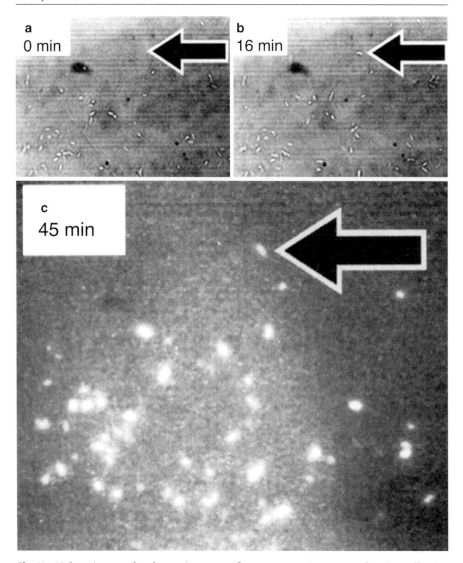

Fig. 22 Light micrograph taken using an epifluorescence microscope, showing adhesion of cells of *P. aeruginosa* to glass surface. This strain (PAO 1) had been cloned to include a *LacZ* reporter construct linked to the promoter of the a*lgC* gene involved in alginate synthesis. **a** The microscope field at time zero. **b** The adhesion of a cell at the location shown by the *arrow* at 16 min. **c** The transcription of the fluorescent product of the *lacZ* reaction at 45 min. The cell was not fluorescent at 31 min, so we deduce that the a*lcC* gene was up-regulated between 14 and 29 min after the cell adhered to the surface. Note that all of the cells that make up the microcolony at 45 min express the a*lgC* gene and appear to have produced large amounts of matrix material. Courtesy David Davies

gration in monospecies cultures could occupy hundreds of graduate students for generations, we should probably "cut to the chase" and examine surface association and cell movement in mixed natural populations because species obviously interact in this process. One dreams of using modern methods of in situ speciation, like FISH probes, to follow the colonization of a cellulose surface in the rumen and to produce a visual sequence of primary colonization and interspecies association that precedes actual biofilm formation. Eventual metabolic cooperation is best served by early associations, prior to matrix production and mature biofilm formation, and the existence of these associations will prove that consortia within these communities develop by signal-based communication rather than by mutual cross-feeding.

The realization that postadhesion behaviors and initial matrix production would require the expression of several specific genes, and Dave Davies' demonstration that a*lgC* is up-regulated upon surface association did not prepare us for the shock we experienced when we first used proteomics to analyze gene expression in biofilms. The initial indication came from Hongwei Yu (Fig. 23a), but the method used was the old-fashioned PAGE gel, and we did not react properly until Karin Sauer unleashed her powerful 2-D gels (Fig. 23b) and showed that the proteins produced (genes expressed) by cells in biofilms differed profoundly from those produced by planktonic cells of the same strain (Sauer and Camper 2001; Sauer et al. 2002; Allegrucci et al. 2006). Hongwei examined the outer membrane proteins of cells in biofilms formed on glass wool within cultures of *P. aeruginosa* and compared them to the OMPs of planktonic cells from the same vessel. Figure 23a shows that biofilm cells produce a set of OMPs that differs almost completely from those produced by their planktonic counterparts; the figure also suggests that the two phenotypes would differ profoundly in their sensitivity to antibiotics. However, Sauer's 2-D gels showed the production of hundreds of different proteins (Fig. 23b) by both biofilm and planktonic cells of many species of bacteria, and the bombshell realization was that these gene products differ by as much as 70% in location and intensity. The cold hard molecular facts are that biofilm cells differ from planktonic cells with the same genotype, because they express a profoundly different set of genes when they are growing in matrix-enclosed communities associated with surfaces or interfaces.

The patterns of gene expression in biofilms and in planktonic populations have now been compared, in many different species and using many different proteomic techniques (Oosthuizen et al. 2002) and gene arrays (Beloin et al. 2004), and they have been found to differ by between 20 and 70%. If we consider that certain "housekeeping" genes must be expressed in any living cell, to sustain protein synthesis and basic physiological processes, these kinds of differences in gene expression would be expected to produce profoundly different cells. Several teams have begun the laborious analysis of exactly which genes are expressed in biofilms and in planktonic cells, and our first impression has been that many of the biofilm-specific genes have been

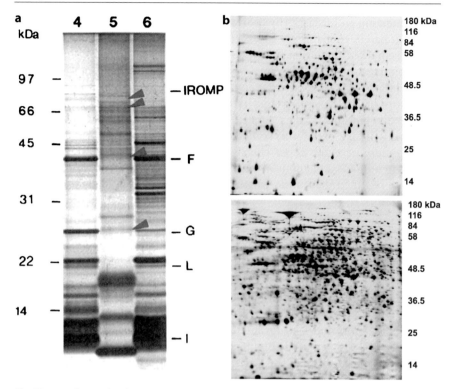

Fig. 23 Protein production can serve as an indicator of gene expression by bacterial cells, and it can be analyzed by simple polyacrylamide gel electrophoresis (PAGE) techniques (**a**) or by the more sensitive 2-D gel electrophoresis (**b**). **a** In 1990 Hongwei Yu recovered biofilms from the surfaces of glass wool in liquid cultures of the PAO 1 strain of *P. aeruginosa*, extracted their outer-membrane proteins (OMPs), and compared them (Lane 5) with OMPs extracted from planktonic cells (Lane 6) of the same strain grown in the fluid of the same culture, and with protein standards (Lane 4). The OMPs of sessile cells differ very profoundly from those of their exact planktonic counterparts. (Courtesy Hongwei Yu). **b** Crude protein extracts were obtained from planktonic cells (*top*) and biofilm cells (*bottom*) from liquid cultures of *S. pneumoniae* and run on 2-D gels by methods outlined in Allegrucci et al. (2006). Note the radical differences in protein production between these two distinct phenotypes. (Courtesy Magee Allegrucci and Karin Sauer)

identified as "unknown ORFs". If we recall that all studies of bacterial gene expression have been conducted using planktonic cells, then we are faced with the exciting probability that many of these genes will produce products that have remained unknown because gene expression in biofilms has never been studied. Because more than 65% of chronic bacterial infections are caused by bacteria growing in biofilms (Costerton et al. 1999), this raises the intriguing possibility that we will discover new pathogenic mechanisms as we study these pathogens in the same phenotype (biofilm) that they are now known to express in infected tissues. One presenter at a recent bacterial genetics meeting enthusiastically proclaimed that > 80% of the ORFs of a strain of *E. coli* had been identified in terms of their gene products, and my biofilm buddy remarked that "that must be one sick lab strain", incapable of either biofilm formation or pathogenesis! In fact 0157 strains of *E. coli* have been shown to lose > 37.5% of their genome (Fux et al. 2005b) on serial transfer in laboratory media. Several groups have begun to analyze which genes are up-regulated in biofilms, using a threefold up-expression as their "gate", and this database will take decades to complete, but we will gradually build up a detailed picture of which genes enable biofilm formation and sustain chronic disease in humans and animals.

As we begin to determine whether biofilm bacteria actually express the genes that make them targets for conventional antibiotics, we find that the penicillin-binding proteins that comprise the target for beta-lactam antibiotics are generally missing in biofilms. This raises the distinct possibility that biofilm bacteria lack the targets for conventional antibiotics, all of which were selected for their efficacy against planktonic bacteria, and that this lack of cognate targets may account for the inherent resistance of sessile bacteria to these agents (Stewart and Costerton 2001). The concept would, therefore, be that conventional antibiotics penetrate biofilms very effectively but that these specific agents fail to find appropriate targets in the enzyme systems produced by the genes expressed in the biofilm phenotype of the organisms concerned. While this concept rationalizes the failure of conventional antibiotics to resolve biofilm infections (Gilbert et al. 2002), the impact of the concept raises more important prospects, in that a careful examination of the genes that are actually expressed in biofilm cells may allow us to design new classes of antibiotics that specifically target sessile bacteria. Both the in vitro and in vivo tests that were used to select all conventional antibiotics used the agent to challenge planktonic cells, and we can provide clinicians with a completely new armamentarium of antibiofilm agents by targeting biofilm bacteria in the same manner. Pharmaceutical companies followed the conventional wisdom of microbiologists and concluded, as early as 1980, that antibiotics to control acute planktonic bacterial infections faced a diminished and complex market, and the "pipelines" are now almost empty. Happily, these companies now realize that new agents that would kill biofilm bacteria preferentially would be very successful, especially if they killed sessile cells of ubiquitous villains like MRSA and/or streptococci, and the search for new classes of biofilm-specific antibiotics has begun. This search can be accelerated to a rate far exceeding the plodding pace of directed synthesis and natural product screening using Sauer's elegant biofilm gels (Sauer and Camper 2001; Sauer et al. 2002; Allegrucci et al. 2006), which reveal exactly which genes are expressed in pathogenic biofilms. These methods are being modified to examine biofilms in vivo, so that new antibiotics can be designed to block critical physiological processes in organisms growing in the phenotype and in the microniche in which they actually cause millions of infections every year.

We will (collectively) examine gene expression in an increasing number of different bacterial species using both proteomic and expressomic methods, but the data already available to us must stimulate us to face some harsh realities, and sooner is much better than later. We have evidence that the biofilm phenotype predominates in nature and in chronic bacterial infections, and we must now examine that phenotype (Ghigo 2001) to determine how different it is from the planktonic phenotype of the same strains in culture, which we have studied as virtual surrogates. In *Pseudomonas aeruginosa* the biofilm phenotype differs from the planktonic phenotype by ±70%, using threefold differences of expression (Sauer et al. 2002) as the "cutoff", so we must face the possibility that significant numbers of the enzymes and toxins we have studied in cultured cells may not be produced in biofilms. If they are not produced in biofilms in vitro, we can determine whether they are produced in commensal or pathogenic biofilms in vivo, and we can determine whether antibodies against them are produced in infected individuals (Brady et al. 2006). If we cannot find these molecules, or any evidence of their presence from indirect clues (antibodies), we are obliged to conclude that they may be significant in acute infections but that they play no role in natural populations or in chronic infections. Brilliant young graduate students are the life blood of our scientific field, and I suggest that it is incumbent on their supervisors to determine whether the enzyme of toxin they are assigned to clone and study is actually produced in the disease at which their research is aimed. Students should not be embarrassed at their posters, or have their first papers rejected by referees and editors, because some of their trusted mentors continue to extrapolate from the planktonic phenotype in culture to the biofilm phenotype in nature and disease.

Common sense must outweigh tradition and convenience, in the choice of phenotype to be studied, and a useful example may be found in Rita Colwell's brilliant and comprehensive studies of *Vibrio cholerae* (Rivera et al. 2001). This aggressive pathogen could hardly be more planktonic, as it scours the natural mucus and commensal biofilms from the human gut, to produce "rice water" stools teeming with mobile and highly infectious bacterial cells. But this sinister organism retreats into the marine environment, between epidemics, and integrates itself into natural microbial biofilm populations, in which it often jettisons the most complex of its pathogenicity islands (e.g., "el Tor") in favor of more rapid growth and better community integration (Schoolnik et al. 2001). When water temperatures reach permissive levels, the sessile cells of *V. cholerae* in natural marine biofilms reacquire the genetic elements that enable them to "go ashore and cause havoc", and they attack vulnerable human populations as planktonic cells in contaminated water sources. Studies of the genetics of the loss and acquisition of pathogenicity islands in *V. cholerae* have been conducted when both the recipients and the donors are planktonic cells, but to be meaningful and ecologically relevant, they must be conducted when the recipient cells are growing in the biofilm phenotype. Bacteria grow in a number of very distinct phenotypes in which gene-expression patterns are profoundly different, and we can no more extrapolate between these phenotypes than a botanist can study pollination by incubating bees with roots and seeds!

Long before the discovery of the biofilm phenotype in 1996, bacteria were known to adopt the spore phenotype and to react to starvation by adopting the dormant "ultramicrobacteria" (UMB) phenotype. These phenotypes were readily recognized because they were morphologically distinct from planktonic cells, but these differences were mediated by the up and down regulation of only a few hundred genes that caused visible differences in cell shape and dormancy. While there are few visible differences between planktonic and biofilm cell shapes, in *P. aeruginosa* these phenotypes differ in the expression of ca. 3000 genes, many of which are involved in community structure. For this reason it is valid to compare single planktonic cells with complex integrated biofilm communities, in which individual cells are connected by pili and nanowires and interact with cells of many other species, and to conclude that these phenotypes differ more than any others. A spore has a cell coat that makes it different from a planktonic cell, and a UMB is much smaller than a planktonic cell, but a biofilm is a multicellular community that differs from a planktonic cell as much as an oak tree differs from an acorn.

1.3.3
Recruitment into Biofilms

While it is tempting to draw distinct boundaries around biofilms, at the outer limits of the matrices that surround the component cells these communities are actually in full communication with the bulk water phase. This communication is enhanced by the formation of open water channels (www.springer.com/978-3-540-68021-5: Movie 1), which are sufficiently capacious to allow the passage (www.springer.com/978-3-540-68021-5: Movie 2) of large eukaryotic cells (e.g., leukocytes) (Fig. 24) (Leid et al. 2002), and are "lined" by the terminal elements of the biofilm matrix fibers in a manner similar to the community boundaries. This open structure allows both prokaryotic and eukaryotic cells to enter biofilms, but they encounter an integrated functional community, and the basic rules of ecology dictate that mature climax communities resist the entry and integration of extraneous organisms. Our direct observations of biofilms in real ecosystems indicate that the recruitment of new cells into biofilms is resisted, as when we flooded the digestive systems of milk-fed calves with cellulolytic bacteria and found that the introduced cells all appeared in the feces. We could, however, integrate cellulolytic organisms into the rumen biofilms of calves that had been weaned (Cheng and Costerton 1986) because this nutrient shift stressed the biofilm community, and stressed communities are more amenable to the recruitment of new members. Our new awareness of the complexity of

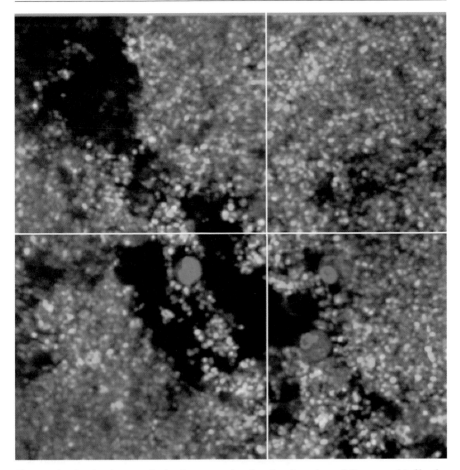

Fig. 24 Confocal micrograph, in the $x-y$ axis, showing invasion of *S. aureus* biofilm by polymorphonuclear leukocytes (PMNs). The bacterial cells have been stained with the BacLite probe, so that live cells are *green* while dead cells are *red*, and living bacterial cells can be seen in the membrane of a PMN near the *center*. The PMNs, whose large nuclei are stained *red*, penetrate 8 to 10 μm into the biofilm and become "paralyzed" in that they retain their membrane integrity, but they are incapable of movement or of killing bacteria. From Leid et al. (2002)

biofilms suggests that membership is established in the earliest stages of community development and that metabolic integration involves the formation of physical and chemical relationships that are much more difficult to forge later in the process. Before these concepts gelled in our minds, we played with a program in which we colonized gnotobiotic ("germ free") lambs with bacterial strains designed to "kick-start" rumen function, only to find that the bacteria acquired from maternal feces did the job must faster and considerably cheaper!

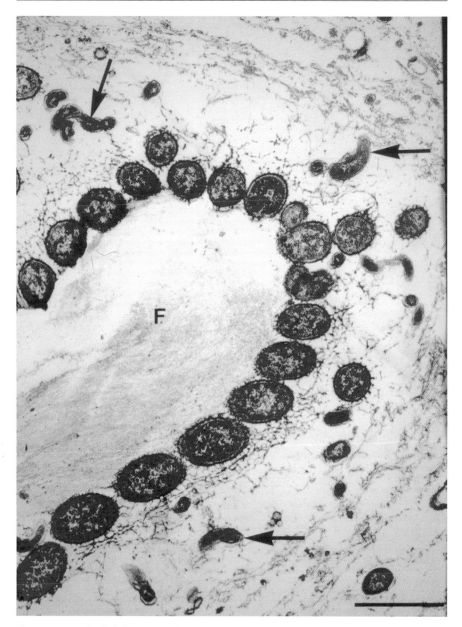

Fig. 25 TEM of cellulolytic microbial population in natural rumen contents. The cellulose fragment (*center*) is colonized by a monospecies biofilm drawn from the mixed population of the rumen. The primary cellulose degraders in the matrix-enclosed biofilm form shallow pits in this nutritive substrate, and they are always associated with spiral Treponema cells (*arrows*) that remove butyrate, which inhibits cellulose digestion if it is allowed to accumulate. The Treponema are highly motile and move to areas of high butyrate concentration by chemotaxis

When K. J. Cheng and I (Cheng et al. 1980) studied the bovine rumen, we noted that the major primary cellulolytic species of bacteria could only digest this substrate at 1/100th of the rate seen in the operating ecosystem if they were incubated with cellulose in vitro. We had noted that digestive biofilms on plant materials in the rumen often included, at their periphery, large numbers of spiral cells of a *Treponema* species (Fig. 25) that had resisted isolation and characterization (Kudo et al. 1987). When we obtained axenic "cultures" of these treponemes and added them to pure cultures of primary cellulolytic species, the rates of digestion shot up to the levels seen in the rumen, and microscopy revealed swarms of berserk corkscrews flitting from place to place like hummingbirds on amphetamines. It transpired that the rates of cellulose digestion were being limited by butyrate, which is a classic end-product inhibitor of cellulolytic enzymes, and the treponemes detected this luscious molecule by chemotaxis and flitted between butyrate banquets. This example illustrates that bacterial cells do not have to be structurally integrated into biofilms, in order to play a vital role in metabolic processes, and that we must consider opportunistic mobile "recruits" to be an essential part of the biofilm community. Scavengers that are guided by chemotaxis are much more efficient if they are not integrated into a structured community, and we expect to find them in increasing numbers of natural biofilms.

1.3.4
Detachment from Biofilms

Biofilms shed cells and components into the bulk fluid by two main processes (www.springer.com/978-3-540-68021-5: Movie 8). The first is a simple sloughing of clusters of cells, and their enveloping matrices, when shear forces overcome the tensile strength of the biofilm. This process was captured digitally by Paul Stoodley (www.springer.com/978-3-540-68021-5: Movie 9), and it is almost certainly responsible for the dissemination of biofilm fragments in natural ecosystems, as well as for the shedding of very dangerous fragments from biofilms in native valve endocarditis (Rupp et al. 2005). The sinister feature of this sloughing process is that the cells in the biofilm fragments are still in the biofilm phenotype, with all of the consequent resistance to antibiotics (Fux et al. 2004) and host defenses, as they are disseminated throughout the circulation. The detachment of planktonic cells from biofilms is a much more complicated process (Purevdorj-Gage et al. 2005) because these sessile cells must both return to the planktonic phenotype and disentangle themselves from matrix components before they can leave the community. This process must include a signal that triggers the synthesis or the release of enzymes that can degrade the basic polymers that constitute the biofilm matrix as well as the conversion of the cells to the planktonic phenotype. In *P. aeruginosa*, the lyase enzyme that digests alginate is continuously synthesized and stored in the periplasmic space (Boyd and Chakrabarty 1994) and is released to digest the matrix when detachment is initiated. Our strong suspicion that detachment is controlled by signals has been supported by Dave Davies' discovery of the detachment signal for *P. aeruginosa* biofilms, and this signal causes such a wholesale detachment event that the previously colonized surfaces are left bare. This end of the spectrum of detachment strategies is invoked in natural biofilms that suffer stagnation, when nutrients are depleted and end products accumulate, and we have seen large areas (e.g., 300 cm^2) of complete biofilm detachment in the hot springs at Yellowstone.

The less extreme day-to-day strategy of biofilm detachment uses the same basic mechanism, at the level of individual microcolonies, in that cells in the centers of the towers and mushrooms that constitute mature biofilms experience stagnation and express detachment signals (Figs. 1 and 11). Because these signals are released in the centers of these microcolonies, the first cells that are seen to break free of the matrix and begin to swim are those in the centers of microcolonies (www.springer.com/978-3-540-68021-5: Movie 11), and a "hollowing out" (Fig. 26 and www.springer.com/978-3-540-68021-5: Movie 10) of the largest of these structures is seen in most biofilms. In some cases the return to the planktonic phenotype gradually spreads from the center of a mushroom, so that the swarm of swimming planktonic cells finds a breach in the dissolving microcolony, and the individual cells swim to freedom as the walls of its erstwhile home collapse (www.springer.com/978-3-540-68021-5: Movie 11). The systematic shedding of planktonic cells begins very soon after biofilm formation is initiated, and large numbers of these free-swimming or floating cells are shed from even the youngest biofilms, which indicates that this is truly a programmed activity of these communities. Because detachment is signal controlled, and because many signal mechanisms are influenced by environmental factors, the suggestion by Bell et al. (1982) that detachment rhythms in natural ecosystems might be species specific and diurnal now seems less heretical and more likely.

If we examine biofilm communities in terms of the mobility options of individual cells and of clonal groups of cells, a much more comprehensive and broad strategy emerges. Cells can assume the planktonic phenotype, which sometimes includes the capability for flagellar motility, but many can also move along surfaces by twitching motility mediated by type IV pili (Mattick 2002), and we simply do not know if cells twitch away from biofilms on surfaces. But we do know that many bacterial species are capable of swarming, which is the coordinated movement of cells along surfaces mediated by pili and lubricated by surfactants, and swarming is just as effective in avoiding stagnation as is detachment. We have tended to see a behavior in a bacterial species, in culture conditions, and then to "tag" that species with an exclusive label as a "swarmer", because cells of that particular species swarm under artificial laboratory conditions. Myxobacteria are the current swarming champions of the microbial world (Pelling et al. 2005); their swarming behavior is signal controlled, and this process results in optimal utilization

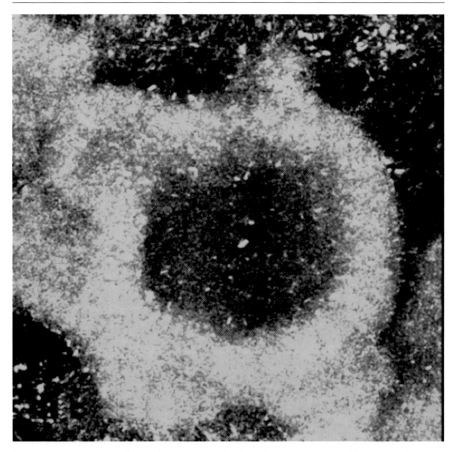

Fig. 26 Light micrograph showing a "hollow" microcolony formed by GFP expressing cells of *P. aeruginosa* on a glass surface in a flow cell. These hollow colonies are often seen in mature biofilms produced by the PAO 1 strain of this organism, and they indicate that some dispersion of the sessile cells has begun. Note that single adherent cells show bright light emission, while the light emission from sessile cells in the microcolony is diffracted and "softened" by the matrix material. See Movie 11. (Courtesy Laura Purevdorj-Gage and Paul Stoodley)

of scattered nutrients in its environment and, eventually, in the formation of special protected fruiting bodies. We propose to take the myxobacteria saga and generalize to mixed bacterial populations, on the basis that the mechanism has been shown to operate in several other species in cultures, and to propose that bacteria in real ecosystems can resort to detachment or twitching or swarming when they need to relocate. In a more limited way, the formation of aerial loops of cells by the marine organisms (Labbate et al. 2004) studied by Kjelleberg's group in Sydney constitutes a dissemination mechanism in that new surfaces are colonized in a saltatory manner resembling the spread of strawberry plants by runners. Cultures have taught us that bacteria can carry out many fascinating activities, and they have shown us the mechanisms by which these feats are accomplished, but how these capabilities operate in the real world remains a mystery that we will unravel by direct observation.

1.4
Resistance of Biofilms to Stress

It is impossible take the long-term ecological view of bacterial invasion of an ecosystem as you apply cold compresses to the brow of a child dying of meningitis or diarrhea, and we have all seen this gut-wrenching battle in terms of bacterial resistance to antibiotics and host factors. We have now completed Phase I of the control of bacterial diseases, by the application of culture-based microbiology to acute epidemic diseases, but the answers that will allow us to complete Phase II of this process will grow directly out of this type of ecological perception. Bacterial recalcitrance in the face of antibiotic therapy (Gilbert et al. 2002) and intact host defenses preoccupies those of us who practice medical microbiology, but this phenomenon is only a part of the amazing ability of biofilm bacteria to survive stress in virtually all ecosystems. Bacteria in the deep oceans and the deep subsurface survive the ultimate stress of complete nutrient deprivation by the Rip van Winkle strategy of the "deep sleep", but the microbial biofilms that dominate all earth surface ecosystems are almost equally adept at survival. The key to stress survival is diversity (Boles et al. 2004). In microbial ecology, as in plant and animal ecology, a single stress can kill all of the cells in a very homogeneous population. But species throughout the biological spectrum actually survive stress by purposely developing phenotypic and even genomic diversity (Ghigo 2001), so that populations facing a single stress contain individuals capable of resisting that challenge. Plants and animals accomplish diversity by mixing and matching genes in sexual reproduction, but bacteria are doomed to the less entertaining alternative of using specialized genes to cause random recombinations in other genes to achieve the same result (Costerton 2004).

1.4.1
Resistance of Biofilms to Antibacterial Agents

Our long-standing preoccupation with planktonic bacteria has allowed us to exercise a large measure of control over diseases caused by these floating and swimming organisms, but it has limited our ability to extend this control to the modern spate of biofilm infections. Thousands of molecules that interfere with the metabolic machinery of bacteria have been discovered in natural sources, or produced by directed synthesis based on physiological information, and the efficacy of these antibacterial agents has been assessed in cultures and in animal models. The antibiotics that we use to combat biofilm bacteria in cystic fibrosis pneumonia (Lam et al. 1980) were selected on the basis of their ability to kill rapidly growing planktonic bacteria in fluid media, or in "lawns" on the surfaces of moist agar plates. It should come as no surprise that these agents are very effective at killing the planktonic bacteria that cause exacerbations in CF pneumonia but have virtually no effect on the recalcitrant biofilms that thrive in the terminal bronchi and eventually destroy the lungs (Doring et al. 2000). These antibiotics advanced in corporate selection processes, and in FDA approval, on the basis of their ability to kill planktonic cells of clinically important "target" species and on the basis of their ability to resolve acute animal infections caused by these free-floating cells. Logic would insist that these "sabot" antibiotics were aimed at the correct enzymes in the bacterial metabolic machinery but that the target bacteria against which they were tested were the minor (planktonic) phenotype in chronic bacterial diseases caused by the alternate (biofilm) phenotype.

The first observations of the inherent resistance of biofilm infections to antibiotic therapy occurred in the 1970s, against a confusing background of the very different and very alarming emergence of bacteria that are resistant to the same agents because of metabolic adaptations. Metabolically resistant bacteria pose a straightforward problem because planktonic cells derived from these infections are demonstrably resistant to the agents concerned, and clinical experience parallels the lab data until some winning combination of agents is found. But biofilm resistance was anomalous, in that culture data indicated the susceptibility of planktonic cells, while clinical observations showed some palliation of symptoms but no resolution of the infections. Resourceful clinicians quickly learned to remove biofilms, on inert medical devices (Khoury et al. 1992) or on compromised tissues, and to use antibiotics to treat acute exacerbations and prevent recurrences caused by planktonic cells. So we entered the 1990s with a partial clinical solution and a continuing enigma. In rural Alberta, corrosion company reps told pipeline operators that their lines had developed "resistant" bacteria, and they urgently needed the more expensive new super-duper biocide to stop early leaks and potential disasters. In the cities, doctors told patients with infected AV shunts and TPN lines that vancomycin would probably solve their problems, but that resistance might develop, and they had to look to Big Pharma for newer and better antibiotics. Meanwhile, biofilms lurked in the shadows, and explanations slowly emerged from the pale weedy geeks in the university on the hill.

When we produced biofilms of corrosion-causing and pathogenic bacteria in the lab, in Alberta and Montana, the first question we asked was "how can all of these cells grow at optimal rates in these complex communities"? Cells near the boundaries of mushroom-shaped microcolonies obviously had prime access to nutrients (Figs. 1 and 11), including oxygen, and they grew and divided rapidly, but the less favored cells in the interior of these communities were also intact and metabolically active. When we observed the effects of draconian stresses, like bleach (NaOCl), we noted that the oxidative wave moved through the community progressively and killed cells and dissolved matrices, without exception, until stoichiometry took over and the agent was depleted. When we observed the attack of more targeted biocides (e.g., quaternary ammonia compounds) and of antibiotics, we noted that some cells always survived single stresses, and that these survivors were distributed throughout the whole community. Kim Lewis later christened these tough cells "persisters" (Lewis 2001; Spoering and Lewis 2001) and wondered aloud why they lived while their neighbors died, but the pivotal fact is that selective antibacterial agents do not kill all of the sessile cells as they diffuse through biofilms, and the survivors are distributed throughout the community. The corollary is that surviving cells find themselves in a perfect nutrient paradise, in a puree of the bodies of their neighbors, and regrowth rates of stressed biofilms are truly phenomenal when the stress is removed.

The well-developed mature biofilm poses a daunting target for any single conventional antibiotic (Lappin-Scott and Costerton 1995; Donlan and Costerton 2002; Jass et al. 2003; Fux et al. 2005a). Any planktonic cells that may be present will quickly succumb, but slow-growing sessile cells will pose a problem for many classes of antibiotics, and anaerobic areas with high proton concentrations may provide a milieu in which certain agents cannot function. Sessile cells with substantial growth rates will have assumed the biofilm phenotype (Drenkard and Ausubel 2002) in which the gene products against which the agent is effective may not be produced, and transport systems that allow access to the cytoplasm in planktonic cells may not be synthesized. Efflux pumps that negate the efficacy of specific antibiotics may be activated in biofilm cells, and the fact that gene expression in sessile cells differs by as much as 70% raises the probability that agents selected for their efficacy against planktonic cells will fail against biofilms. Phil Stewart has drawn on his unique background in chemical engineering and diffusion theory to analyze the inherent resistance of biofilms to antimicrobial agents in a masterful review that considers all of these points in a balanced and dispassionate manner (Stewart 1996). In addition to the physiological variability inherent in communities that contain fast and slow-growing cells, engaged in both aerobic and anaerobic metabolic processes, Pradeep Singh has shown us that specific recombinant genes are up-regulated in biofilms (Boles et al. 2004). This adds programmed genomic diversity to the background of metabolic diversity, so that no two cells in a biofilm are truly identical, and the "job" of any single antibiotic or combination of conventional antibiotics is made almost impossible. Perhaps we should marvel at the observed fact that all of the sessile cells in the thick biofilms (Fig. 27) built on native heart valves by viridans group streptococci (Mills et al. 1984) can eventually be killed by high-dose, long-term (6- to 8-week) therapy with conventional antibiotics. But then we return to the real world, with ICU patients showing pus and inflammation at the exit sites of five different "lines", and we wonder what can be done to prevent the incursions of environmental biofilm formers into the human ecosystem.

The present crisis in medical microbiology and infectious disease involves two types of bacterial resistance to antibiotics. In the first type, specific pathogens mutate the genes controlling either the metabolic target of the antibiotic or the transport systems that allow access of the agent to the target, and this property is then disseminated to other members of the same or other species by horizontal gene transfer that often involves plasmids. The resistant organisms (e.g., methicillin-resistant *Staphylococcus aureus* or MRSA) represent the epitome of bacterial adaptation, and they thrive in unprotected patients and proliferate in hospital environments. The problem is acute, the mechanism is well understood, and the solution lies in better asepsis and in the accelerated development and controlled deployment of new classes of agents. The second type of resistance is less well understood, and it involves the functional resistance of bacterial populations in the device-related and other chronic bacterial infections that have gradually come to predominate in modern medicine (Costerton et al. 1999). The mechanism of this resistance is complex and devolves from the basic characteristics of biofilms, in that all cells in the pathogenic populations have adopted the biofilm phenotype and in that individual cells differ radically in metabolic activity and even in genotype. These resistant organisms already exist in all areas of the hospital environment, they affect virtually all patients compromised by instrumentation or by underlying physiological compromises, and they represent the epitome of the bacterial strategy of survival by diversity. When the bacteria combine their two basic strategies and we are faced with metabolically resistant organisms living in phenotypically and genotypically diverse biofilms, we face our sternest challenge, and we need all of the weapons of modern science to succeed.

Logic leads us to two very different strategies to control chronic biofilm infections. In the first, we will simply replace planktonic cells with biofilm cells in the screening of potential antibiotics, and we will bow to conventional wisdom by producing these sessile communities in natural body fluids under natural body conditions. Both George O'Toole's 96 well plates (O'Toole and Kolter 1998) and the Calgary EMBEC system (Ceri et al. 1999) provide practical ways of exposing biofilms to new antibiofilm agents, and killing efficacy can now be measured by several culture-independent methods. My own calculations of the probability of success of various agents has been abysmally disappointing, and so I favor an empirical approach that screens very high numbers of antibiotics and combinations of antibiotics and ancillary agents until in vitro success is fully validated. These agents can be advanced toward clinical applicability by the use of realistic animal models, but we can also take advantage of the slow progress of most human biofilm infections to monitor efficacy by direct observations of specimens from clinical treatment protocols. The second strategy is much more intellectually satisfying, in that

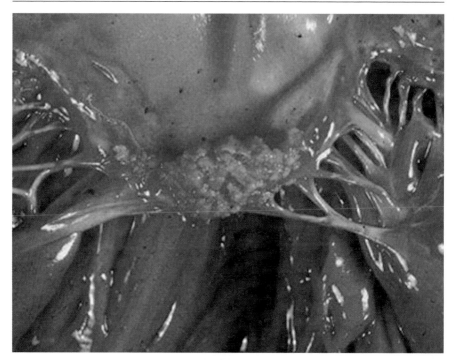

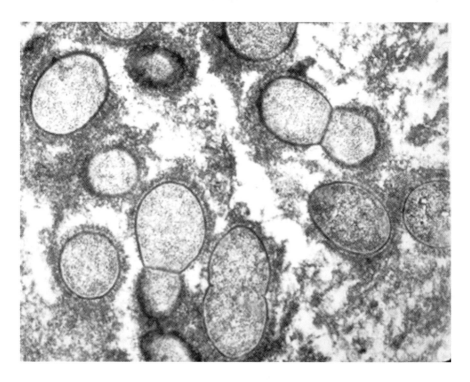

◄ **Fig. 27** *Top*: photograph of a "vegetation" formed on a human tricuspid valve by infecting cells of viridans group Streptococci. *Bottom*: TEM of a similar vegetation formed on the tricuspid valve of a rabbit in an experimental infection by the same organism. Note the very extensive matrix surrounding these Gram-positive cells, which often contains very large amounts of host-derived material, including damaged platelets

the capability of biofilm formation can be denied to the invading organisms, so that they are left "swinging in the wind" [sic] and in the highly vulnerable planktonic phenotype. Biofilm blockers have been developed, by mechanisms discussed below (Sect. 4.2.3), and these agents already show bright promise in the prevention and treatment of chronic biofilm infections. Our chances of controlling biofilm infections increase with every increment in our knowledge of the causative community and with every little morsel of information about the genes and molecules involved in the construction of these microbial citadels (Parsek and Fuqua 2004)!

1.4.2
Resistance of Biofilms to Environmental Stress

The development of resistance to whole classes of clinically (over)used antibiotics is only a recent, and very minor, "blip" in a 3-billion-year process that has given bacterial communities the stress resistance that enables their dominance in even the most extreme environments. This property allows both biofilm cells and planktonic organisms to survive the concerted attack of metabolic poisons, whose natural analogs they will have encountered many times during the interspecies warfare that rages continuously in natural ecosystems. To understand the remarkable resilience of bacteria, we need to ponder their ability to colonize and persist in ultraharsh natural systems, as well as their ability to lurk in the hospital ecosystem while the whole medical and cleaning staff is bent on their destruction. The two basic concepts that illuminate the resistance of bacteria to environmental stress are that the community is the unit that operates in evolution and that the driving force is the "ambition" of these communities to colonize all available surfaces in all permissive ecosystems. The population pressure that drives barley terraces up the precipitous slopes of the great mountains of the Himalayas, until tiny level patches host a few plants in the shadows of glaciers, also drives the microbial biofilms that creep toward boiling hot springs and menacing fungal conurbations. If all of the planktonic "scouts" dispatched by a microbial community perish because of a particular stress, then the community itself can spread along the surface, like mediaeval soldiers under their interlocked shields, and probe the utmost limits of colonization. While humans and other sentient creatures need many years to extend their territory, bacterial communities can lose a surface to a particular stress one day, and reoccupy it in a few hours of the next day, if the stress is relaxed.

When Canadian scientists examined the threat of airborne bacterial attacks in the "heat" of the cold war, they discovered that drying and ultraviolet (UV) light killed planktonic cultured cells in seconds. Because these same stresses operate in the tidal zones of oceans, we know that bacterial communities survive periodic dryness and UV light by encasing themselves in matrix material and then recruit their photosynthetic cousins to turn their precarious perches to their advantage. Boiling water kills all bacterial cells except spores, but burgeoning biofilms lend an unearthly beauty to Morning Glory pool in Yellowstone Park, as microbial communities adapt to this steady heat stress and obtain exclusive possession of this extreme ecosystem by pressing the limits of stress survival. Antarctica tests the mettle of all living creatures, but bacterial crusts thrive in the outer shells of boulders in the dry valleys (Staley and Konopka 1985), and huge masses of bacterial biofilm have been found in the depths of Lake Vostok (Christner et al. 2001), whose surface has been sealed by ice for millions of years. Extremes of heat cold and dryness have earned these ecosystems their human designation as "extreme environments", but the bacterial communities that live in tidal flats, hot springs, and the dry and moist ecosystems of Antarctica belie this designation by growing luxuriantly and "living large". In these environments, any planktonic cells that wandered out of the perimeter of its microbial community would suffer the same fate as that of a besotted oil worker who staggered out of Tuktoyaktuk without his parka or his rifle.

While these extreme examples show the upper limits of the adaptation of microbial communities to simple physical stresses, the success of these communities in more salubrious but more competitive ecosystems offers just as many insights. In soil, which may be the epitome of a complex ecosystem, surfaces are plentiful but simple nutrients are soon consumed, and dominance in the lower profiles must be ceded to communities that can cooperate to process complex nutrients in toxic anaerobic environments (Crawford et al. 1977). Nitrification is balanced with denitrification and cellulose decomposition, and the most desperate threadbare communities will eke out a miserable living by chewing away at tannins and lignins until only oil and gas are left in the lowest reaches. These "farming" communities must remain sedentary among their complex nutrients, because there is little energy to spare, but highly social and very mobile myxobacterial communities make regular forays through soil before attracting human attention by forming beautiful macroscopic fruiting bodies. These myxobacteria can lead us toward an accelerated understanding of microbial biofilms because they have retained many of their behaviors during the dark ages of culture-based microbiology, and they can serve as an example of many behaviors that may be common in these communities.

Membership in a myxobacterial community enables chemotactic gliding cells of individual species to move among soil particles, with their sedentary colonies of primary producers, in a socially coordinated manner that maximizes uptake of readily available nutrients. Like the Cossacks of old, the mobile myxobacteria can pillage accumulated nutrients while riding stirrup to stirrup with cells of their own or related species, and chatting back and forth by means of specific vesicle-enclosed signals (see Sect. 2.3). The sedentary microbial communities of soil famously produce pungent antibiotics for protection from marauders, and drying must affect the mobile myxobacteria more than the slime-enclosed primary producers, so the mobile community will eventually feel stress. The resources of the myxobacterial community come into play when the stress is "perceived", and some community members are designated for altruistic sacrifice in that they die and dissolve to provide nutrients for the construction of the fruiting bodies where designated cells can form cysts and survive (Kaiser 2004). We happen to know a lot about the myxobacteria, because their macroscopic fruiting bodies have yielded pure cultures of individual species, but there are certainly many other predatory bacteria in soils and other natural ecosystems. Predation carries inherent dangers in that the prey may produce antibiotics, squadrons of predators can become separated if signals fail, and mobile predators are more sensitive to negative environmental factors (e.g., drying), so diversity is beneficial if it allows some individuals to survive and thrive.

Meanwhile, back in the sedentary communities that predominant in most environments, stresses come from many sources and the community is sufficiently committed to its special nutrient source that relocation is not an option, unless the nutrient is exhausted or an alternate source is available. Predation is a constant stress, but the matrix serves to exclude bacterial predators (e.g., Bdellovibrio) (Kadouri and O'Toole 2005) and to allow survival during phagocytosis by amoebae (Murga et al. 2001), while growth in the biofilm phenotype may produce fewer cues for predatory chemotaxis. Planktonic scouts are susceptible to predation, but their demise does not compromise the survival of the community. The community provides very effective protection from nutrient stress, in that individual cells contain all of the nutrients necessary for survival, and diversity will dictate that some cells will die and release their contents in time to enable the survival of other cells. The biofilm matrix retains these nutrient molecules and traps more from the bulk fluid of the environment, so that very little is lost from the community, and bacteria can always reduce their metabolic activities to conserve energy. The spatial organization of many primary communities enhances their nutrient sufficiency, as in the case of cellulolytic communities in the bovine rumen, in which the actual cutting edge of the community on the insoluble substrate is comprised of "robo-vesicles" (Fig. 20) packed with enzymes, but lacking DNA. In the event that a particular member species is lost to the community, perhaps by its susceptibility to a physical stress (e.g., drying), that member can be rerecruited from the genome pool of the ecosystem when conditions return to optimal.

Stress by nontargeted chemical agents, such as acids and surfactants, can be tolerated by a microbial community if some cells (e.g., acid-tolerant mycobacteria) survive and reconstitute the consortium that exploits the nutrient opportunities of the ecological niche. The binding properties of the matrix are pivotal in community survival because the molecules released by dying cells are retained for recycling, and DNA fragments of genes of importance to the whole community are retained for future uptake and use. Targeted antibacterial agents exert a profound stress on all microbial communities, in competitive ecosystems, and their survival depends directly on their diversity. One natural antibiotic (penicillin) depends on cell-wall synthesis as its specific target, so cells that are not actively growing are unaffected, and virtually all mature biofilms contain some quiescent cells. Most targeted antibacterial agents require a channel or a pump to facilitate access to their specific target inside the cell, so cells that can deny that access by mutation of their cytoplasmic membrane components become resistant to the antibiotic. High rates of horizontal gene transfer (Fig. 10) and accelerated recombination rates give biofilms a very high level of genomic diversity, and this "plurality" makes it more likely that a small number of cells in any microbial community (Boles et al. 2004) will survive the attack of any single antibiotic.

While all of the properties we attribute to biofilm communities seem to contribute to stress survival, and while the arguments I have marshaled to make this connection may make sense, the proof of their success in the face of multiple stresses trumps all speculations. The stresses are very real, but microbial biofilm communities dominate the biosphere and constitute its largest biomasss, so we must conclude that they cope very effectively indeed! The last passenger pigeon has flown, and the mountain pine beetle has virtually doomed the pine forests of western North America, but bacterial communities assemble like clockwork and thrive whenever a new nutrient opportunity presents itself anywhere in the biosphere. The ultimate survival strategy of biofilm communities may be the vast repository of bacterial genomes in the deep sea and deep subsurface and the "hard-wired" programming in these genomes that allows them to assemble multispecies consortia for any nutrient opportunity. Even if Earth wobbles in space, and the biosphere as we know it is fried to a crisp, newly emerged black smokers in the sea floor will still be colonized by biofilms of sulfide- and methane-oxidizing bacteria that will live and thrive "on the edge" near columns of toxic steam. Those of us in the medical business must think very hard if we are to outmaneuver this very old and very successful bacterial life form, and perhaps learn to speak their language, and even enlist them in our never-ending fight against disease.

1.5
Biofilms as Opportunistic Self-Mobilizing Communities

When we first noticed the spatial juxtaposition of metabolically cooperative bacteria (Figs. 1 and 25) in the very efficient consortia that carry out virtually all of the degradation of complex insoluble nutrients in natural ecosystems, microbial ecology was in its infancy and our speculations were simple and immature. The notion that bacteria are attracted when a neighbor produces a nutrient substrate of interest involves the simple concept of chemotaxis, and the notion of the rapid growth of the fortunate cells that ended up in the preferred location is satisfying, if simple-minded. This concept of spatial association of mobile bacteria, driven by nutrient advantage and for the purpose of metabolic efficiency, still lingers in the minds of microbial ecologists to this day. A parallel (and ridiculous) concept in the eukaryotic area would be that mesenchymal cells associate with the notochord in the developing embryo because one produces some organic acid to which its partner is attracted, and that both cell types burgeon when they meet because they are well fed! But this simple concept of nutrient cooperation matched the evolutionary level and degree of sophistication that was generally accorded to bacteria, which were on the very lowest rung of the organized biological entities.

Two new perceptions challenge us to rethink the developmental sequence that produces metabolically integrated multicellular communities and offer us the possibility of positing a more accurate and more useful "embryology" of a functional multispecies biofilm. First, we discovered that starved bacterial cells are converted to very small dormant "ultramicrobacteria" (UMB) (Fig. 4) that retain their full genomic complement (Kjelleberg 1993) and resuscitate to full size and full metabolic activity when nutrients again become available. UMB derived from virtually every bacterial species that has ever lived on earth predominate in the deep ocean and in the deep subsurface (Balkwill et al. 1997), and certain numbers of these dormant prokaryotes are present in virtually every ecosystem in which nutrient content varies between feast and famine. Therefore, in all natural ecosystems, a vast library of genomes is available from which individual genomes can be mobilized and mixed and matched to produce multispecies communities that are ideally suited to capitalize on any nutrient opportunities that are presented. So when a fish dies in a freshwater ecosystem or a black smoker emerges in a marine ecosystem, the individual bacterial genomes that can evolve into a custom-made rapid reaction biofilm are ubiquitous and readily mobilized (Fig. 28 and www.springer.com/978-3-540-68021-5: Movie 12). When a biofilm population develops in response to the availability of nutrients, it constitutes a nutrient source for other heterotrophic bacteria, and complex communities develop at the favored site (Fig. 29 and www.springer.com/978-3-540-68021-5: Movie 12). Unlike many higher organisms (e.g., insects) that must be present in an ecosystem in a suitable form to be able to take advantage of an episode of nutrient availability, hundreds of bacterial genomes are omnipresent and capable of immediately capitalizing on all such opportunities (Fig. 30 and www.springer.com/978-3-540-68021-5: Movie 12).

Secondly, we have discovered that bacteria produce chemical signals that they use to control intraspecies behavior patterns (including biofilm forma-

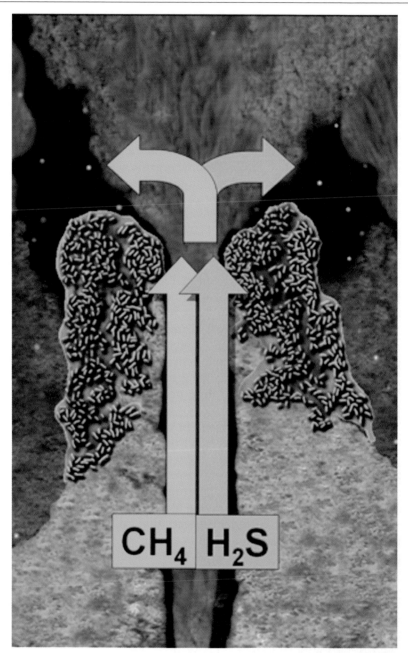

Fig. 28 Conceptual drawing of the emergence of a black smoker that releases methane and hydrogen sulfide into an environment containing UMB of species (*blue* and *green*) capable of oxidizing these energy-rich compounds to support cell growth. Biofilms composed of vegetative cells of the methane oxidizing (*green*) and hydrogen sulfide oxidizing (*blue*) species will develop in immediate proximity to the newly opened vent

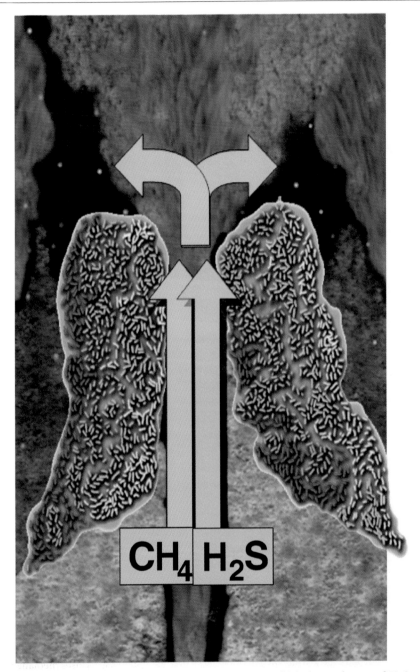

Fig. 29 Conceptual drawing of a mature black smoker in which the methane and H_2S oxidizing biofilms serve as organic substrates for heterotrophic species (*purple* and *white*) that form complex biofilms in association with primary colonizers. Eukaryotic organisms then form communities in which the whole bacterial ecosystem serves as a nutrient base

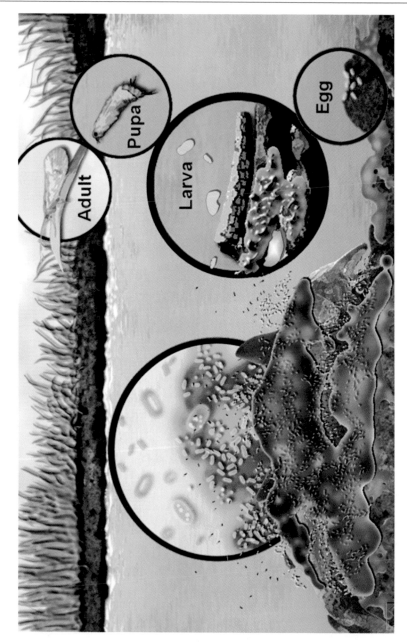

Fig. 30 Conceptual drawing of a dead fish in a stream, in which the unfortunate creature provides nutrients for a wide variety of bacteria that form extensive biofilms on its surface in order to take full advantage of this gastronomic opportunity. Insects can share the bounty if they happen to be present in the larval form, by consuming the fish or the bacterial biofilms, but they cannot benefit from this largess if they are in the egg, pupa, or adult stages of their development

tion) and that also mediate interactions between cells of different bacterial species (Xie et al. 2000), and even interactions with eukaryotic cells including those of host organisms. These well-defined signal systems constitute a possible mechanism for the control of a "hard-wired" development strategy, somewhat similar to the process whereby the sequential firing of hormone genes controls the complex processes of embryological development in higher organisms. Many contemporary reductionist studies have examined the association of two or more bacterial species in the early stages of biofilm formation, and a sufficient number of signal-based interactions have been discovered to prepare our minds for a more sweeping concept. We can imagine that the microbial populations that develop and are sustained on tissue surfaces (e.g., lactobacilli on the epithelium of the human vagina) originate from genomes resident in the system, and produce the same types of population in most cases, because the process is guided by a collective "firing pattern" of the genes that comprise these genomes. In this schema an environmental trigger, like the sudden availability of a certain type of nutrients, would attract certain primary and secondary colonizers onto adjacent surfaces, in a pattern that would be dictated by preprogrammed signal interactions and that might or might not confer any immediate nutrient advantage to the participants (Fig. 31). This notion would raise the evolutionary position of bacteria from a primitive level in which each individual cell behaves in its own best interests to one where each cell within a biofilm is an integral component of a complex multicellular "organism" in which the success of the individual is subservient to the success of the whole community (Caldwell and Costerton 1996).

Bacteria are unique in the ubiquity of their genomes and in their ability very rapidly to assemble (Fig. 28) these genomes into microbial communities that can derive energy from the most dilute and the most transitory of nutrient opportunities. Because even the most oligotrophic aquatic environments contain a rich variety of ultramicrobacteria (UMB), thousands of bacterial genomes are omnipresent, and these dormant bacteria can resuscitate and initiate community building in a matter of minutes (www.springer.com/978-3-540-68021-5: Movie 12). These metabolically integrated communities can process small concentrations of an easily degraded organic molecule very rapidly and then return to the planktonic phenotype and undergo starvation survival to form a new crop of UMB. When insoluble and refractory organic molecules enter an aquatic ecosystem, UMB of species with the appropriate enzymes will resuscitate and form multispecies biofilms that will gradually process the nutrient and distribute the resultant energy within their consortia. So the bacterial strategy of starvation survival provides universal genomes, while the other bacterial strategy of biofilm formation assembles communities, and bacteria can benefit from all nutrient opportunities however small and however transitory. This opportunistic pattern of bacterial growth has allowed these organisms to dominate the vast ecosystems of the

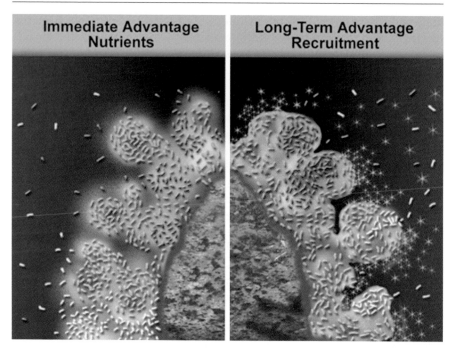

Fig. 31 Conceptual drawing contrasting colonization strategies. In the first strategy (*left*) the four member species cluster in microcolonies because of immediate nutrient advantages, in that one or more species metabolize the end products of other species, and they construct functioning consortia; In the second strategy (*right*) two of the four species (*blue* and *green*) emit signals that attract two other species (*purple* and *white*) to form integrated metabolic consortia for no immediate, but considerable long-term, mutual benefit

deep ocean and the deep subsurface and to operate successfully in extreme ecosystems that cannot support any other life forms.

Bacterial nutrient strategies are equally successful in nutrient-rich ecosystems because of their ubiquity and speed of reaction and because microbial communities process nutrients at all stages in their development. If a fish dies in a stream, the unfortunate creature represents a nutrient source for all inhabitants of that ecosystem, but microbial biofilms will harvest the lion's share, because their chief competitors must be present in a nutrient-responsive state to share in the banquet (Fig. 30). Insects must gain access to the food while they are in the larval or adult stage of development, and the bacteria often dine alone if flowing water or the coffin lid excludes all competitors. Bacteria are perhaps most successful if they form prokaryotic/eukaryotic communities that cycle a constant source of nutrients, like the sunlight impinging on a microbial mat, and the wonderful variety of bacterial genomes can always supply the species that fits perfectly into each niche in these metabolically integrated communities. Because bacteria comprise the major component of the biosphere, and because the majority of these bacteria live in biofilms (Kolter and Losick 1998), we can submit in evidence the fact that this strategy of living in highly reactive self-assembling communities has served these organisms very well indeed!

1.6
Efficiency of Biofilms

1.6.1
Physiological Efficiency of Biofilms

When we summon up our collective understanding of microbial physiology, we visualize a planktonic cell that acquires organic nutrients from the rich "soup" of a special medium and oxidizes these molecules to yield energy that is stored by extruding protons (Mitchell and Moyle 1965). These energy reserves can be "cashed in" when these protons are allowed to reenter through the cytoplasmic membrane, where they generate ATP and other phosphorylated nucleotides that drive virtually all of the anabolic processes of the cell. This elegant arrangement functions well in dense suspensions of planktonic cells in rich media, like those we might find in the silage and sauerkraut that become acidic because of this proton extrusion, but it will not work at all well in most natural ecosystems. Individual bacterial cells in turbulent oligotrophic ecosystems, like freshwater streams and the bulk water phase of rivers and oceans, would not remain in contact with the protons they extrude and could not therefore recover their investment in energy. Furthermore, the very scarce organic molecules on which they depend for nutrients would only be encountered momentarily, in a whirling vortex, in a manner that would not favor uptake by the membrane's fine-tuned transport enzymes. One of the most important factors in explaining the remarkable predominance of biofilms in nature may be the physiological efficiency of these matrix-enclosed multicellular communities (Fig. 1). Organic nutrients are delivered into biofilms by the convective flow of bulk water through well-defined water channels (www.springer.com/978-3-540-68021-5: Movie 1), and these molecules then partition into the matrix surrounding the sessile biofilms cells, where they remain available for transport into the cells. We depend on this nutrient trapping and storage by biofilms, on a practical level, when we "feed" our diminutive subjects for as few as 1 hour in 24 and obtain excellent growth even though the biofilms only see nutrients for very limited periods of time.

This ability to trap and store nutrients may be pivotal in turbulent systems like mountain streams, where organic nutrients are scarce and contact between molecules and cells in the bulk fluid must certainly be fleeting, at best. The provision of nutrients in these systems is episodic, and the death of a fish would provide few benefits to planktonic bacteria passing in the water, while the nutrients from the vertebrate disaster can (and do) nourish billions of bacteria that live in the biofilms that soon form right on the nutrient source. The basic strategy of biofilm formation immediately on nutrient sources can readily be seen near sewage outfalls in streams and rivers, where it is obvious that sustained sources stimulate the formation of macroscopic biofilms that trap the nutrients and recycle them as biomass. Stable juxtaposition with other nutrient sources, like photosynthetic algae, also favors sessile cells in biofilms, and specific multicellular communities are seen to form immediately next to other energy sources like the methane and hydrogen sulfides of "black smokers" in the marine environment (www.springer.com/978-3-540-68021-5: Movie 12).

More subtle nutrient opportunities are provided to sessile bacteria in biofilms, when cells of different species set up metabolically cooperative consortia (Fig. 1), as in the case of the mixed bacterial communities (Kudo et al. 1987) that combine to degrade cellulose (Fig. 25). Because insoluble organic molecules (like cellulose) comprise a very large proportion of the nutrient content of natural freshwater ecosystems, and because the microbial degradation of such compounds is carried out by cooperative microbial communities, the bacteria in these systems must form biofilms as a matter of physiological necessity. In these cooperative consortia, individual bacteria may take up nutrients produced by their immediate neighbors, and they can attain very high levels of metabolic efficiency if their own products are removed by cells of other species and they can avoid the physiological perils of feedback inhibition (Fig. 25). In the whole integrated community every cell that extrudes protons, as a consequence of its oxidation of an organic molecule, contributes to a pool of these ions (Fig. 32) in the matrix-filled spaces between the cells. Because the generally anionic polymers that comprise the matrix of biofilms bind cations, including very large numbers of Mg^{++} and Ca^{++} ions, this area surrounding each sessile cell acts like a proton sink and prevents the loss of protons into the bulk fluid (Fig. 32). A real mixed-species biofilm in a real natural environment is a very sophisticated and highly integrated physiological "machine" whose inherent efficiency goes a long way toward explaining the universal predominance of this mode of growth in virtually all nutrient-sufficient ecosystems. Figure 1 goes some distance toward capturing this concept in that the dynamic "kelp bed" of the biofilm captures nutrients that are processed by bacterial consortia in metabolically integrated microcolonies to produce a rich supply of extruded protons that drive all activities in the community.

Perhaps because of the disconnect between the biological sciences and physics, we have failed to include the electrons at cathodically charged metal surfaces in the menu of potential energy sources available to support bacterial growth. This energy is already in the form of electrons and does not require autotrophic metabolism of metal ions, and scientists (notably Ken Nealson of USC) in the exciting new field of geomicrobiology have taught

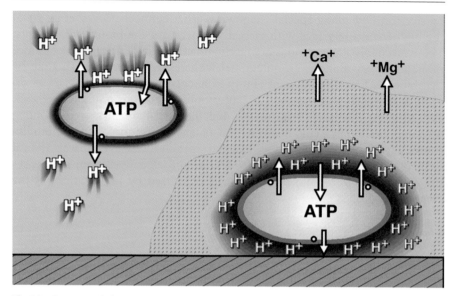

Fig. 32 Conceptual drawing comparing proton extrusion in planktonic and sessile cells. The planktonic cell at left extrudes protons and generates ATP by returning them to the cell but risks losing these ions if diffusion carries them away. The sessile cell at *right* extrudes protons into the matrix of its biofilm, displacing divalent cations in the process, and retains these energy-rich ions in an acid zone surrounding the cell even in high-flow ecosystems

us that biofilms can both extract and contribute energy at metal surfaces (Nealson 1997). When biofilms generate energy to the metal surface that they colonize, we have a fuel cell, but many natural biofilms receive energy from metal surfaces in natural ecosystems and use it to drive what we have heretofore thought of as heterotrophic metabolism. The champion metal-colonizing bacteria are those of the genus *Shewanella* (Caccavo et al. 1992), and these cells both produce metal-binding proteins that dictate their associations with metal surfaces and cytochromelike proteins that manage the reciprocal flow of electrons. Biofilm communities are ideally suited to harvest the huge energy reserves in reduced metals because they colonize metal surfaces and modify the immediate environment of the metal by blanketing it with their matrices. In this cozy association, large amounts of energy are made available to the cells that find themselves immediately juxtaposed to the metal surfaces, but observations and calculations indicate that cells hundreds of microns distant, but in the same biofilm, also benefit from this largess. Dianne Newman has shown (Newman and Banfield 2002) that energy can be transmitted throughout biofilms by the use of special shuttle molecules that can be reduced at the high-energy interface and then can drive metabolic activity by being oxidized anywhere in the same biofilm. Derek Lovely and Yuri Gorby have very recently shown that this energy can also be transmitted by electron flow along special "nanowires" that connect (Gorby et al. 2006) the high-energy interface with virtually all parts of the biofilm community. These mechanistic revelations are fascinating, but the profound "take home" message is that biofilms are integrated with respect to energy. This means that biofilms have the inherent ability to harvest energy by processes as diverse as heterotrophic metabolism, autotrophic metabolism, photosynthesis, and stripping electrons from metals and then to share this energy among all component cells. This coordination of energy resources in the biofilm community takes microbial multicellular communities a long way toward eukaryotic cells, in their levels of integration and cooperation, and raises the question of how this coordination is organized by the genomes of the component species.

1.6.2
Genetic Efficiency of Biofilms

When Cam Wyndham studied the dissemination of pollutant-degrading genes in natural aquatic ecosystems, he found that the horizontal interspecies transfer of these genetic elements occurred 1000 times more rapidly in stream biofilms than in planktonic lake populations (Wyndham et al. 1994). Both somatic and plasmid-born genes (Ghigo 2001) are transferred very efficiently between cells of different species in biofilms in natural aquatic ecosystems, and it appears that these sessile communities may serve as "party pads" for the exchange of resistance plasmids in hospital biofilm communities. Biofilms provide an optimal milieu for horizontal gene transfer because the cells are juxtaposed in a stable matrix (Fig. 10) containing many pili (including F pili), which obviates the necessity of competent planktonic cells finding each other as they wheel through fluid space propelled by Brownian motion. If individual cells in biofilms are positioned by a network of pili, as suggested above (Sect. 1.2), it is easy to visualize the close apposition of donor and recipient cells in a virtual orgy (www.springer.com/978-3-540-68021-5: Movie 3) of horizontal gene transfer within biofilm communities. Because clonal cells of the same species are often seen to comprise individual microcolonies, or to predominate in mixed-species microcolonies, cells of the same species may be ideally situated for the exchange of both plasmids and somatic genetic material.

If we apply Garth Ehrlich's brilliant distributed genome hypothesis (Shen et al. 2005; Ehrlich et al. 2005) to biofilms, we can conceive of a situation in which the complex and genetically expensive machinery required for the degradation of certain complex pollutants could be jettisoned by cells intent on degrading more amenable nutrients. However, as long as at least some cells have retained all of the genetic elements of the degradative pathway, the community as a whole will be ready to mobilize the genome from its distributed sources and swing into action if this complex substrate should suddenly become available. Similarly, human pathogens would not need to maintain energy-expensive pathogenicity islands when growing in biofilms in competitive ecosystems, but they could mobilize these islands when pathogenic opportunities arose, as long as some cells had retained the genetic information. Pradeep Singh uses the same concept (Boles et al. 2004) in elegant studies in which he shows that horizontal gene transfer maintains genetic diversity when cells of *P. aeruginosa* grow in biofilms. He has identified genes that facilitate this process, and he suggests that this diversity allows populations of this organism to adapt to challenges posed by host defenses and by antibiotic therapy as they grow in the lungs of cystic fibrosis patients. We know that the principles of microbial ecology have penetrated into medicine when Pradeep, a brilliant physician, invokes the "insurance" hypothesis first developed in grassland ecology (Hooper and Vitousek 1997) to explain how the genetic diversity generated by horizontal gene transfer can benefit a human pulmonary pathogen.

Our large and very impressive community of microbial geneticists has, of necessity, built up its enormous and very valuable repository of knowledge of the mechanics of DNA replication and gene expression using planktonic cells in single-species cultures. This "hard-nosed" genetic approach, which has recently come to dominate research in microbial physiology, can make equally useful inroads into microbial ecology by the simple absorption of three new but well-documented concepts. Biofilm populations predominate in nature and in modern medicine, and they are structured in a manner that facilitates extraordinary levels of horizontal gene transfer, which is reinforced by specific genes that are switched on only in biofilms. This enhanced horizontal gene transfer in biofilms allows some clones within certain species to jettison certain "high maintenance" genes, not required for current activities, and to reacquire these genes from their clonal partners at a later time and in different circumstances. The phenotypic expression of the bacterial genome changes profoundly when planktonic bacteria adhere to surfaces and form biofilms, and the nature of this expression is controlled by environmental factors, some of which involve metabolic partners and host tissues. If these three concepts can be inculcated, the very effective weapon of microbial genetics can be aimed at real natural ecosystems and at chronic bacterial diseases, instead of being aimed at pure cultures and diseases over which we already have a measure of control. The study of the genetic "hard-wiring" that controls the development of whole microbial communities will be challenging, but potentially very satisfying and highly relevant.

1.6.3
Ecological Efficiency of Biofilms

When Cam Wyndham and his intrepid crew braved strong currents and minus 40 °C temperatures to study bitumen degradation in the Athabasca River, they laid the ecological foundation for an industry that currently delivers 1.3

million barrels of oil each day. Ecological issues had been raised because it had been reported that "grab" samples of Athabasca water showed no capability to degrade oil or bitumen, and government scenarios had visualized an ecological disaster stretching from the nascent oil recovery operations to the Artic Ocean. Cam and his crew showed that 99.99% of the bacteria that had assembled to feast on the hydrocarbons of the tar sands, many millennia before Sun Oil started up their bucket wheels, grow in biofilms on all available surfaces in the river. In some tributaries that flow through rich bitumen deposits Cam saw grey biofilms on the surfaces of black tar on the bottoms of pools, and he found similar biofilms on all surfaces in the main river, except for those of sand bars, which ground up the oil-degrading communities by their constant abrasive movement. The demonstrable consequences of the efficiency of this quite remarkable biofilm community are that all traces of bitumen hydrocarbons have vanished from its sediments by the time the river reaches its delta at Lake Chippewa, just 38 miles downstream from the last tar sand deposits. It was equally gratifying to note that the oil-hungry biofilms of the Athabasca made such short work of 30 000 barrels of synthetic crude that Sun Oil inadvertently spilled when a barge loader went to town for a beer that no trace of the oil could be found on the following day.

When we formed thick (> 3 mm) biofilms on the surfaces of bitumen-soaked concrete blocks in the Athabasca, we noted that these communities were riddled with holes from insect larvae that graze on benthic biofilms and then emerge to bedevil Northern humans. We note that the thick biofilms that form on trickling filters in sewage-treatment plants also provide sustenance for insect larvae, and direct observations of biofilms on rocks in the pristine Bow River show that they sustain very large populations of grazing amoebae (Fig. 6). Because of their remarkable efficiency in digesting solid substrates, trapping dissolved nutrients, and generating organic compounds by photosynthesis and by autotrophic processes, biofilms occupy the bottom rung on the food chain in virtually all aquatic ecosystems. It is ironic that engineers, with typically clear and immediate objectives, developed commercial technologies in which they use biofilms to degrade wastes and to synthesize commodity chemicals, while the mainstream of microbiology was mired in the study of planktonic cells in cultures. The engineers didn't worry about what species of bacteria were present in their systems, or which genes controlled the enzymes carried out the reactions, but they developed iterative models and managed these important processes with little if any help from microbiology.

It is useful, in the light of the obvious success of the microbial communities that predominate in most ecosystems in the biosphere, to ponder the reasons for this resounding triumph. Bacteria are unique in their ability to adapt to starvation by forming UMB that preserve their genomes and persist for very long periods of time in nutrient-deprived environments (Fig. 4). They are equally unique in their ability to rapidly mobilize appropriate genomes (www.springer.com/978-3-540-68021-5: Movie 12) into custom-made communities that can respond to nutrient opportunities (Fig. 28), including those offered by cathodic metal surfaces from which electrons can be harvested. When biofilms form on insoluble nutrient surfaces, digestive enzymes are concentrated in the community matrix (Fig. 20), metabolic partners are held in close juxtaposition (Fig. 25), and the protons that accrue from this communal activity are retained in the community (Fig. 32) for general use. Inequalities in energy within the community are balanced by reduced "shuttle" compounds and/or nanowires, cells can be positioned optimally by a community "skeleton", and the whole community sheds planktonic cells of its component species to respond to opportunities downstream. The functioning community (Fig. 1) has programmed genomic diversity to withstand specific antibacterial challenges, and it is inherently resistant to drying and UV light, but it is susceptible to grazing by higher organisms and serves as the basis of food webs in most aquatic ecosystems. In essence, bacteria are hard-wired to function as members of integrated microbial communities, and we are only just beginning to discover the nature of this highly evolved genetically mediated strategy.

1.7
Relationship of Conventional Single-Species Cultures to Natural Biofilm Populations

In the gradual transition between microbiological research using pure cultures of single species toward the direct study of natural mixed-species populations in situ, it is useful to consider exactly what a pure culture is, and to do this as dispassionately as possible. When a mixture of cells is recovered from a natural mixed population (Fig. 19), aggregates are dispersed as much as possible, and the resultant suspension is spread on the surface of an agar plate, so that single cells or small groups of cells will give rise to individual colonies. This is essentially a clonal process, in that the single cells and the small aggregates are (by definition) clones, and the resulting culture will contain a uniform suspension of genomically identical cells, until random mutations introduce some variations. These clonal cultures have enabled virtually all of the physiological and genetic studies of bacteria to date because, until the advent of in situ methods like reporter constructs (Davies and Geesey 1995), this uniformity was essential when measuring the processes or properties of millions of cells en masse. The disadvantage of these clonal cultures is that they only capture one of the dozens or hundreds of clones of that particular species that live and thrive together (Ehrlich et al. 2005) in the natural ecosystem being studied. When John Govan and others (Nelson et al. 1990) have carefully collected thousands of isolates of *P. aeruginosa* from the lungs of cystic fibrosis patients, they have found hundreds of different phenotypes suggesting the presence of at least as many genotypes, and a lively debate rages as to which is the "most typical".

The challenge of studying a real ecosystem using clonal cultures is further complicated by the phenomenon of genetic drift on subculture. As we remove a drop of fluid from a fluid culture, we select in favor of planktonic cells and against the sessile biofilm cells growing on surfaces within the test tube (Fig. 19). As we remove cells at any particular stage of culture growth, we select in favor of cells that will be alive and ready to propagate and against clones that may have flourished and died during the early stages of logarithmic growth. This ritual act of subculture exerts selective pressure on the clone in question, and all of us who have cultured *P. aeruginosa* from cystic fibrosis patients have watched as five to six serial subcultures transformed a sticky green mass of biofilm to a turbid grey suspension of planktonic cells. Christoph Fux has reviewed (Fux et al. 2005b) this process of genomic drift during subculture, and he quotes published data that indicate that 10 to 15 subcultures routinely transform wild 0157 strains of *E. coli* to laboratory strains that have actually lost 37.5% of their genome. This loss of genetic material during subculture is especially invidious in studies of disease because growth in vitro favors "housekeeping" genes and discriminates against genes needed for pathogenesis and for integration into multispecies biofilm communities. Certainly all genes involved in adhesion to surfaces would be lost on subculture because this phenotype will remain on the walls of the test tube while the "magic drop" is transferred. It is very sobering to realize that gene chips made with type strains of certain pathogenic species, like the PAO 1 strain of *P. aeruginosa* and the K 12 strain of *E. coli*, may not contain many genes that are pivotal in disease processes caused by these organisms. There is a general movement in favor of using clonal cultures that as are close as possible to wild strains of the organism concerned, but we should still remember that the clone we happen to have in the test tube is only one of many clones in the actual ecosystem. In addition, in spite of their current cachet, molecular methods of detecting gene expression are only as good as the cultures that were used to supply the type species DNA, and many widely used arrays were made using very tired old lab strains that couldn't cause disease in a starved diabetic rabbit.

As so often happens, intellectual order is being restored in a chaotic field by a well-trained and unusually perceptive person, and Garth Ehrlich has ridden in on his white horse and proclaimed the distributed genome hypothesis (Shen et al. 2005; Ehrlich et al. 2005). This hypothesis states that any given clone is unlikely to contain all of the genes that can be present in cells of a certain species and that we can only deduce its true genome if we "pool" all of the genes we can find in multiple isolates from a situation in which a number of clones have been successful. In pathogens success might be seen as a thriving infection, and numerical predominance might be an index of success in an environmental situation, but our main premise is that many clones cooperate in most microbial enterprises and that this genetic diversity conveys a huge advantage on the species in question. We live in a wonderful time in which accurate perception is often followed, almost immediately, by the technical means of examining these perceptions. We will be able to examine the concept of genetic plasticity by the use of the two-photon microdissection microscope, recently developed by Zeiss, because we will be able physically to recover as few as ten bacterial cells from pathogenic and natural ecosystems. The DNA from these cells can then be amplified by Roger Lasken's multiple displacement amplification technique (Raghunathan et al. 2005) to yield their complete genome, which can then be compared with the genomes of other clones of the same species, as identified by a common 16 S rRNA sequence and detected by FISH probing. This marriage of direct observation and molecular methods will allow us to determine the genetic makeup of an unlimited number of different clones of the same species, in situ, and to identify which clones thrive in association with certain neighbors or cause specific reactions in the infected host.

If we seize this perception of the reductionist approach that has brought us to the place where we stand today, in microbiology, we may be able to return to the main biological fold (Costerton 2004) by belatedly adopting the modus operandi of our colleagues in botany and zoology. If we mimic our colleagues by simply looking at our living subjects in their natural habitats, using the fabulous new tools for direct observation, we can observe microbial communities (Fig. 1) much as our colleagues might observe a moose or a fir tree. Using transmission electron microscopy (TEM) we can see that real bacteria in the bovine rumen have cell envelopes of virtually baroque complexity (Figs. 7 and 20), while every rumen species we have cultivated has the basic Gram-negative "train tracks" or the Gram-positive "nutshell". Using scanning electron microscopy (SEM) we can see that certain strains of *Staphylococcus epidermidis* retain the ability to construct tertiary structures of remarkable complexity (Figs. 14–17), while most isolates of this skin-associated bacterium grow as suspensions of planktonic cells. Using confocal microscopy, with Mickey Wagner's autoradiography techniques, we can see that cells of one bacterial species can adopt a wide range of growth patterns that depend on their spatial relationships to specific partners in functioning consortia (Figs. 1, 10, and 13). The prospect of what we can do if we take the nuts and bolts of the basic bacterial physiology and genetics we have learned in the past few decades and apply this knowledge to the examination of structures and processes in real microbial ecosystems is very exciting and likely to bring microbiology forward with a giant leap.

In every bacterial genome that has been mapped to date, there are almost as many ORFs of unknown function as there are genes whose product and purpose are known. Every time we look at real bacteria in real ecosystems, by any method of direct observation, we see structures we have never seen before and we detect complex functions that can only be explained on the basis of the coordinated activity of multispecies consortia (Kudo et al. 1987;

Palmer et al. 2003). We need to turn the basic pattern of microbiological research around by 180 degrees. We need to follow the lead of the gallant little band of microbial ecologists, who are perhaps the real biologists among us, and we need to observe functioning microbial populations in situ using microscopy and direct measurements of local chemical parameters. We need to begin to trust direct observations of structures and processes, no matter how complex they are and how much they depart from conventional wisdom, and we are obliged to wrestle with them until we understand them in terms of the nuts and bolts of our culture-derived hypotheses. In this manner we can emulate our colleagues in "myxobacteriology", who have observed the amazing behaviors of these fascinating organisms, and dissected the remarkable structures that they make, and have accounted for each behavior and structure in terms of the genes that are involved. Gliding motility has been explained as the result of retractable pili, swarming has been rationalized as a signal-directed process (Kim et al. 1992), and the mysteries of fruiting body formation has been plumbed in terms of signals and environmental triggers (Kaiser 2004). This fruitful marriage of direct observation and molecular analysis can be an example to the whole field.

Major perceptual advances are often preceded by glimpses of things to come, and the few glimpses we have seen in the past 5 years presage an era in which bacteria will occupy a niche in the biological hierarchy that vastly exceeds their present assignment (Hall-Stoodley et al. 2004). We have noted that all bacteria send and respond to chemical signals, like the marine Vibrios, and I believe that we will soon conclude that all bacteria exhibit social behavior like the myxobacteria. Sometimes it just takes us 35 years to detect general patterns. We have established that bacteria live predominantly in biofilms attached to surfaces, and we now have glimpses of structures that facilitate horizontal gene transfer and energy transfer in these remarkably integrated communities. We have discovered that biofilm bacteria can form elaborate towers and mushrooms, under the control of chemical signal mechanisms, and we now see that they can also construct walls and partitions to condition their immediate environment. These bacteria can control the deposition of walls and partitions of very uniform thickness, at repeating intervals, and they demonstrate a type of tissue-forming capability similar to that of eukaryotic organisms. Functional consortia of Byzantine complexity are found in such humble ecosystems as soil and sewage, and it appears that the availability of certain nutrients, under certain environmental conditions, triggers the assembly of dozens of species for a common purpose (Fig. 1). Our minds are frequently boggled by what we see, but we must find the genes that control these amazing structures and behaviors, and, as soon as our minds are cleared, we must make more direct observations and rationalize them in molecular terms. The fresh young faces we see in Microbiology 101 belong to people who will be very busy, for many decades, in bringing microbiology into line with the other biological sciences and in rationalizing bacterial behaviors in molecular terms.

1.8
Biofilm-Based Understanding of Natural and Engineered Ecosystems

Our initial discovery that biofilms predominate in oligotrophic mountain streams (Fig. 2) was based on the enumeration of bacterial cells, but was actually triggered by gross [sic] observations of the clear slime that covered all surfaces in these systems. These observations were repeated in natural aquatic systems of increasing nutrient content, culminating in abattoir effluents, and the numerical dominance of the sessile population held true throughout this process. We then used "heterotrophic potential" measurements to show (Wyndham and Costerton 1981) that biofilm populations accomplish > 99.9% of the turnover of organic substrates in these ecosystems, so that the use of planktonic "grab" samples to estimate the ability of the ecosystem to "process" organic nutrients became counterintuitive. The hypolimnion at the air–water interface has now been added to the equation, as a de facto "surface" on which biofilms form, and it is generally accepted that the bulk of bacterial transformations that occur in the biosphere take place in these sessile microbial communities.

This preponderance of biofilm bacteria is also seen in industrial water systems, where the tradition of analyzing "grab" samples of the flowing water has led to equally misleading conclusions. For several decades, until the 1980s, the relative risk of microbially influenced corrosion (MIC) was assessed on the basic of the detection of sulfate-reducing bacteria (SRB) in "grab" samples from pipelines and other susceptible installations. More importantly, the success of biocide treatments in controlling these planktonic SRB was monitored on the basis of reduced planktonic counts in the bulk fluid, and treatment was discontinued when they were reduced below a specific level. We now know that MIC is actually accomplished by biofilm communities (Fig. 3), containing SRB and many other electron-shuttling organisms, and that these sessile stationary populations are much more resistant to biocides than their planktonic counterparts. As the biofilm concept became established in the oil industry, biofilm coupons gradually replaced "grab" samples in monitoring, and the corrosion control was based on measures designed to kill bacteria in the biofilms that actually induce corrosion by setting up a classic "corrosion cell" in affected metals. Once again, science caught up with common sense, and the observation by old oilfield hands that "pigging" pipelines with scrapers and high local concentrations of biocides was vindicated and rationalized. The scrapers remove biofilms from colonized surfaces, the biocides kill bacteria that are suddenly made planktonic by the scrapers, and the whole cycle is repeated once a week to keep the biofilms "off balance" and incapable of initiating an attack on the metal. On the outsides of pipelines, MIC is kept at bay by the imposition of cathodic protection, in which D.C. electric fields override any corrosion current produced by biofilm bacteria, and we bury millions of tons of metal in the ground and protect it from bacterial attack with mechanical and electrical wizardry.

The immense strides made by engineers in the past two centuries were made possible, in part, by the development of the "systems approach", in which general principles are established and verified and are then used to predict outcomes in similar circumstances. We submit that highly organized, metabolically integrated bacterial communities (Fig. 1) will form on surfaces in any natural or engineered aquatic ecosystem and that these biofilms will predominate both numerically and functionally in these systems. Sys-

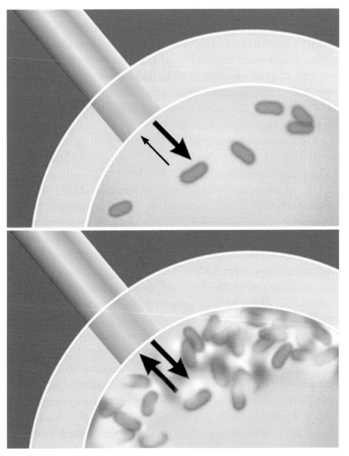

Fig. 33 Conceptual drawing of the Bioptic Biofilm Probe developed by Intelligent Optical Systems (Torrance, CA). The fiberoptic probe is integrated into the system to be monitored, and it delivers UV light at 404 nm into the fluids immediately adjacent to the wall of the pipe or vessel. All bacteria contain NADH and NADPH that autofluoresce (at 432 nm) when excited by the UV light (at 404 nm), and small numbers of planktonic cells (*top*) return very little of this fluorescence to the probe. However, large numbers of stationary sessile cells (*bottom*) return very large amounts of fluorescence, and biofilms can be monitored in a quantitative manner

tem performance is often based on biofilm effects, like efficiency losses in heat exchangers and other cooling systems, and system threats like biofouling and corrosion are often connected to biofilms, so we have concentrated on biofilm detection. Because recovery-and-culture methods are slow and ill-suited to biofilms, we have examined immediate physical measurements of biofilm detection and have settled on the BiOptic probe that provides real time on line biofilm monitoring. This technology (Fig. 33) depends on the autofluorescence of two common bacterial coenzymes (NAD and NADP), when illuminated with ultraviolet light (404 nm), and biofilm cells immobilized on the end of the optical fiber that carries the light to the system return a signal at 432 nm if biofilms have formed. This BiOptic monitor is now commercially available for biofilm monitoring in any and all natural and engineered aquatic systems. The systems approach is equally applicable in any aquatic system, and we depend on its central principles in predicting the preponderance of biofilms in systems as disparate as product packaging lines and coral reefs. Moreover, as our understanding of biofilms improves, we can begin to make predictions in such related areas as species diversity and resistance to antibacterial agents.

1.9
The Evolution of Biofilms

As we prepared the fifth in our series of biofilm reviews for Annual Reviews of Microbiology (Stoodely et al. 2002), I was startled by the appearance of a sweaty and disheveled Paul Stoodley in my elegant director's office at the Center for Biofilm Engineering. Paul had experienced a "road to Damascus" revelation during his headlong run up some nameless Montana mountain, and he feared that he might lose some elements of this inspiration in the time it would take to shower and change to less pungent clothing. In essence, Paul suggested that biofilms may have predominated in the primitive earth and that the more elaborate refinements of planktonic cells may have developed much later in the evolutionary process. He drew parallels to the evolution of plants, in which simple mosses and bryozoans colonized the primitive earth, and the elegant dissemination methods of seed dispersal developed millions of years later, when interspecies competition replaced simple survival as the primary selective force.

The first bacterial cells would have evolved in an oligotrophic environment, in which organic compounds would be very rare, and the nutrient-trapping capability of biofilm communities would favor success and predominance just as it does in contemporary alpine streams. The intermittent streams of the primitive Earth would subject their bacterial populations to severe drying, and biofilms would protect their component cells from dehydration and UV light much like the plant cells in mosses are protected from these stresses. It is unlikely that the stream in the primitive Earth consisted of a sequence of salubrious environments, but it is probable that permissive pools alternated with acidic or hypersaline pools in a volcanic landscape. In such an ecosystem, the ability of a community to maintain itself in a permissive location would be *de rigueur*, and the biofilm mode of growth would again be favored. Biofilms would predominate in these ancient extreme environments (Krumbein et al. 2003), as they do in harsh contemporary ecosystems, and the lives of most of the planktonic cells released from these communities would be "nasty, brutish, and short" (Thomas Hobbes). The notion that the bacterial life forms that succeeded in the primitive earth may persist in extreme environments in our current terrestrial state offers the intriguing prospect of using these ecosystems for retrospective thinking on a scale that beggars the imagination.

As the Earth matured and streams evolved into a series of nutrient-rich environmentally permissive ecosystems, interspecies competition would replace simple survival as the imperative. In this case, the ability of a planktonic cell to move rapidly and to follow a chemical gradient to locate favorable organic nutrients would provide a distinct ecological advantage and justify a significant investment in genetic capacity and energy. The fascinating and elaborate molecular mechanisms that mediate gradient sensing, and motility in response to these gradients, would then acquire a selective advantage, and organisms that produced planktonic cells with these attributes could preferentially colonize downstream niches. As animals began to appear in this prokaryotic domain the bacteria would have taken refuge in biofilms to survive the attack of amoebae, and this "experience" would "stand them in good stead" when they later undertook to invade multicellular animals and faced phagocytic white blood cells. At the zenith of the natural association of bacteria with humans, before the development of antibacterial agents in the past 200 years, some microbes had evolved the very successful strategy of lurking in protected biofilms and attacking with highly evolved planktonic cells. These bacteria exploited human ecosystems, without wiping out this nutrient-rich niche, by attacking with planktonic "missiles" that homed in on specific tissue targets and killed selected individuals before they could mount an acquired immune response. Paul's notion that bacteria first evolved a biofilm strategy for survival in challenging environments and later developed specialized mechanisms of colonizing particular niches (including the human body) offers us a useful perception. As we have gradually made life much more difficult for specialized human pathogens, they may have reverted to their basic survival strategy, hunkered down in biofilms, and adopted the same survival modus that they used in more primitive days. Bacteria have thrived in the biosphere, while larger species have waxed and waned, by alternating between a basic biofilm phenotype that produces stable integrated communities and a planktonic phenotype that produces less protected cells capable of remarkable sophistication and aggression.

2   Control of all Biofilm Strategies and Behaviours

The intellectual challenge facing all microbiologists, and all biologists who study systems with prokaryotic components, is to grasp and embrace the unequivocal and well-supported concept that bacteria and fungi (Douglas 2003) live in complex multicellular communities. This sweeping paradigm shift demands a resolve to discard the ingrained mental image of swirling clouds of planktonic cells and the imagination to sketch in a picture of integrated communities (Ghannoum and O'Toole 2004) whose borders are still ill-defined but are roughly located by certain salient facts. The mental picture that we struggle to imagine must accommodate the facts that cells of one *P. aeruginosa* clone form mushroom "stalks", at whose apices a subpopulation undergoes programmed apoptosis to provide a template for the formation of the mushroom "cap" by cells of another clone that climbs the stalk by twitching motility. We must visualize aggregates of myxobacteria that swarm across surfaces, in a highly coordinated manner, and then sacrifice a similar subpopulation to provide a template for the spiral towers that become fruiting bodies. Our image must accommodate the observation that plants can summon nitrogen-fixing partners that form biofilms on roots, penetrate the host tissues, and set up highly structured metabolically integrated intracellular communities that contribute to plant function. The new picture must be elastic enough to incorporate Shewanella biofilms within which dozens of specific binding proteins mobilize iron ions and then transmit the resultant energy to all parts of the multicellular community using redox shuttles and very long protein "nanowires". We must absorb the fact that the epithelium of the rumen of the newborn calf uses signals to "select" a biofilm population that will detoxify urea and scavenge oxygen and then responds to the presence of these microbial partners by completing its differentiation as an organ system. If the particles in an anaerobic digester consolidate their structure, so that a methanogenic core is surrounded by concentric layers of heterotrophs with graded hydrogen tolerance, we must be open to the notion that other microbial communities undergo dynamic restructuring.

The reasons that we are obliged to stretch our thought processes in these agonizing extensions, whether we are ready or not, is that these bacterial "behaviors" have been documented by reliable colleagues within the strict rubric of scientific communication. If we cannot disprove these reports, we cannot dismiss them as charming hobbitlike tales, but we must determine which genes direct them and look for homologs throughout the Prokaryotic Kingdom. We must also look outside the test tube and see if these remarkable behaviors are exhibited in nature by organisms that feign innocence and Amish-like simplicity when we trap them and grow them in glass prisons. It is unique in the history of biology that the practitioners of one discipline (microbiology) have taken 160 years to discover the basic mode of growth of their living subjects, and now the only factor that limits the explosive burst that will follow is our ability to imagine new complexities.

2.1
The Mobilization of Biofilm Communities

As we struggle to wrap our minds around the complexities of mammals, with their finely tuned molecular cycles, we can take some comfort from the inexorable workings of genetics as individuals reproduce and their progeny develop by repeated embryological processes. Even with the Zebra fish and the humble toad, we can watch the embryological expression of a single genome that will produce hundreds of individuals with similar phenotypes, with a healthy dash of diversity provided by the mixing and matching of genes. When bacteria appeared to live as simple swarms of individual prokaryotic cells, their genetic continuity appeared to involve simple redistributions of the finite number of genes among individual sister cells. But now we face the daunting challenge of adapting these ideas to explain the development of highly structured metabolically integrated communities (Fig. 1) of different species of bacteria, which reproduce themselves with the same regularity seen in higher "monogenome" organisms.

Bacterial communities are rapidly mobilized in constrained ecosystems with direct contact with similar communities in the same ecosystem, as when newborn calves mobilize rumen populations very quickly ($<$ 4 d) from maternal sources. More commonly, bacterial communities are mobilized from a large number of individual genomes (www.springer.com/978-3-540-68021-5: Movie 12), and this process is "directed" by the chemical and physical parameters operative in the ecosystem in question. If we think of each species as having a "distributed genome" (Shen et al. 2005; Ehrlich et al. 2005), which is the compendium of all genes contained in large numbers of individuals of that species, we can conceive of a community as a collection of such comprehensive genomes. If we consider the well-documented complexity of various microbial communities (Figs. 13, 25, and 29), we must conclude that the comprehensive genomes of certain bacterial species are "programmed" to interact in the formation and (eventually) in the function of communities tailored to specific ecosystems. While the intricate choreography of monospecies embryology juxtaposes cells and tissues that must combine to make the individual organism, the mobilization of any microbial community demands that cells of each species find each other and make functional connections. Just as neurons in the developing mammalian brain make connections with cells with similar phenotypes, bacterial cells in developing biofilms must make connections with cells with similar and dissimilar phenotypes to construct an integrated community of somewhat lesser sophistication. This demands that the comprehensive genome of the bacterial species we find in integrated communities must be coordinated, and programmed for cooperation, in a pattern that is new to biological thinking.

In higher multicellular organisms we observe the interactions of cells with identical genomes but different patterns of gene expression, and we see the development of individuals similar to their parents. In prokaryotic multicellular communities we observe the interactions of cells with different genomes, even within the same species, but we see that these programmed associations result in a structured and integrated "organism", albeit one of lesser complexity and sophistication. We already see evidence of interactions between bacterial species from the same biofilm community, as when cells of *Streptococcus mutants* stimulate the growth of cells of *Lactobacillus acidophilus* in a dental biofilm, but the study of these interactions is still in its infancy. We get titillating glimpses of the mechanisms that bacteria use in the building of multispecies communities when we culture cells of different species together in liquid media, or when we grow them as adjacent streaks on agar, but the data are still too scattered to discern a pattern. Attempts to reconstruct the interactions required for community development a priori, from studies of mono- and multispecies biofilm preparations, will only produce a nightmare of bioinformatic gridlock. The interactions are simply too numerous and too interdependent. I submit that the most productive approach will be to observe physical juxtaposition and functional cooperation between different species in real biofilm communities (Fig. 25) and then to examine the spatial and functional relationships between these organisms throughout the developmental cycle.

Microbial communities undergo a developmental process (O'Toole et al. 2000) involving the proliferation, position, and integration of cells that is controlled by a variable number of genomes. For this reason, it may be instructive to examine some representative communities to explore the upper limits of the structural and functional complexity that can be achieved by multicellular organisms that are assembled from numerous comprehensive prokaryotic genomes. In marine sediments, we can see the formation of extensive ($> 10 \text{ mm}^2$) sulfur-oxidizing "veil" communities (Thar and Kühl 2002) within which certain members rotate their flagella while anchored to the veil structure and thus provide the sulfur-oxidizing members with sufficient oxygen for their task. In soil we can watch with fascination while many different myxobacterial cells join in mobile aggregates that use their retractable pili to glide over solid surfaces, in social packs, and respond to suboptimal conditions by making and mounting spiral ramps to form fruiting bodies. All of these social activities of particular myxobacterial species are carried out against a background of other bacterial species, including other myxobacteria, and yet the single-species functional unit is maintained. Other bacterial species play a role in multispecies communities, like that of the bovine rumen, in which cellulose and many other polymers are digested while essential physiological processes (e.g., protein digestion, urea reduction) are maintained. It is not my purpose to examine these complex communities in detail, but only to call attention to the "landmarks" that may lie along the borders of the capability of microbial communities to approach the level of eukaryotic multicellular organisms in complexity and sophistication.

To help us wean ourselves from the concepts we have absorbed in 160 years of looking at bacteria in single-species cultures, perhaps it would be useful to imagine a single species, or maybe a genus, with all of the attributes we now acknowledge to exist in some specialized organisms. Let us imagine a heterotroph that is among the primary colonizers in a photosynthesis-driven mat community that produces methane in its lower anaerobic regions, and let us give our species genes that direct it to associate with specific photosynthetic and methanogenic species. Let us then give our species the ability to synthesize pili and nanowires that connect it to its metabolic partners for precise positioning in the community, for horizontal gene transfer, and for energy sharing. Let us then confer on our species the ability to synthesize and organize a protein scaffold in the interstices of the community matrix on which a mobile partner can be persuaded by diffusible signals to settle and "ventilate" the community with its flagella. Let us then give our species genes to attract hypermobile *Treponema* to approach the biofilm community and clean up any end products that might accumulate as a result of its coordinated metabolic activity. Let us then confer on our species the ability to produce planktonic cells that can glide, using retractible pili, and swarm in coordinated packs along chemical gradients that lead them to similar operating communities in the same ecosystem. Let us assist this migration by having our species produce vesicles full of signals that are specifically addressed only to cells of this same species, so that the swarms can stay together as they move from one community to another (perhaps more successful) one. Let us confer genes that promote horizontal gene transfer, and resultant genomic diversity, so that some of the company will always survive if the photosynthetic partners get buried in sediment, and the community is forced to survive on trapped organic compounds. In case the worst should happen, let us give our species the ability to swarm to the edges of the biofilm and construct arboreal structures so that selected cells can climb into the branches and form spores where they have some chance of dispersal. When confronting one of these cells, in a single-species culture, in a rich medium, she would hotly deny having any such nocturnal adventures or special abilities, and she would claim to be simply "helping scientists with their enquiries".

2.1.1
Signal Gradients in Microbial Biofilm Communities

The initial discovery (Hastings and Nealson 1977) of signal communication between bacterial cells occurred in the late 1960s, against the background of the microbiological concepts of that era, and their designation as "quorum sensing" molecules (Fuqua et al. 1994) reflects this timing and these concepts. In liquid cultures, and in the light organ of the bobtail squid, the bulk fluid concentration of signal molecules was proportional to the number of cells of the producing species, and the wide-ranging effects of these signals adapted bacterial behavior to cell numbers. The planktonic cells did not attempt to light the fires of Lucifer until there were enough of them to be effective, and bacteria in the circulation of an animal did not start to produce toxins (Passador et al. 1993) until there were enough of them to do some damage. The logic is impeccable, and signals certainly regulate planktonic populations in that way, but most bacteria in most ecosystems grow in biofilms. How do signals regulate the formation, structure, and function of biofilms? We can deduce, from this treatise up to this point, that cells within biofilms display exquisite control of their positions and their activities, and we suspect signal control, but we cannot simply transfer the planktonic model of quorum sensing into the biofilm rubric.

The perception that microbial communities develop by a series of orchestrated processes, somewhat analogous to the embryological processes orchestrated by a single genome in multicellular eukaryotes, leads me to consider the coordination options available to prokaryotic cells in this situation. While the functional form of a mature multicellular eukaryote is produced and coordinated by the sequential expression of genes that produce hormones, the development of multicellular multispecies prokaryotic communities is controlled by a symphony of cell–cell signaling (Pesci et al. 1999; Fuqua and Greenberg 2002). Because bacteria have finely tuned patterns of chemical signaling in response to changes in their chemical and physical environments, we can readily understand how a primary colonizer can share its "joy" at finding its favorite substrate with others of the same and related species. This first burst of diffusing signals would serve to recruit potential members of the climax community because cells of the same species would work together (Fig. 1) to initiate biofilm formation, and cells of metabolically allied species would begin to congregate. Then, as in the eukaryotic equivalent, the programmed response of cells and groups of cells to signal gradients would control a process whereby the cells of all the component species would form structures to optimize the metabolic processes for which the community was assembled. The gradual formation and structural evolution of "poppy seed" granules in methanogenic wastewater reactors provides a case in point. Soon after the wastewater is placed in the system, soft grey-brown aggregates appear in which the heterotrophs that convert organic compounds to organic acids are mixed in interlocking patterns with the methanogenic archea that convert these acids to methane. Then, over the next 3 weeks, the granules become smaller, harder, and darker (eventually black), and we find a core of methanogenic archea that receives organic acids from a mantle of heterotrophs and churns out methane bubbles at a rate hundreds of times higher. Methanogenic wastewater reactors are excellent examples of microbial systems that were developed and optimized by engineers, based on the work of a few isolated microbial ecologists, long before the notion of integrated microbial communities entered the mainstream of microbiology.

Research in the signal control of community development is in its infancy, and it is hobbled by the fact that signals were discovered in studies of planktonic bacteria and have only very recently (1998) been expanded to studies of single-species biofilms (Davies et al. 1998). The discovery of acyl homoserine lactone (AHL) signals of Gram-negative bacteria (Fuqua et al. 1994), the cyclic polypeptide signals of Gram-positive bacteria (Dunny and Leonard 1997; Balaban et al. 1998), and the exciting autoinducer II signals of the whole kingdom (Schauder et al. 2001) have spurred a gold rush for signal inhibitors. But this frantic search yields few scientific dividends because the search is conducted in a manner that only detects the complete inhibition of biofilm formation; the signals and cognate inhibitors that affect community development are not detected. We speculate that, while there are signals whose pivotal position in the signal network causes them to influence whole processes like biofilm formation, there will be many more signals that control some aspect of biofilm architecture (e.g., water channel dimensions) very specifically. If we think of the subtle control of individual cells in biofilms, from coordinated twitching to produce aggregates that will become microcolonies to the release of detachment signals by deeply buried cells in stagnant niches, then we must recognize full analogy with the hormones of higher organisms.

If we take a naïve and simplistic view of cell–cell signaling and adapt it to the control of processes within microbial communities, we can visualize small molecules that react with cognate receptors and activate DNA transcription by these dual-purpose proteins (Fig. 34). If the signal controlled the activity of some vital component of the EPS synthesis cascade, it would stimulate the production of this matrix material (Fig. 35, top) and initiate biofilm formation, and there are literally hundreds of cellular activities that are known to be controlled in this way. While the development and function of microbial communities is controlled by the production of these hormonelike signals by some cells and by their specific effects on other cells of the same or different species, a significant level of control can also be exercised by signal inhibitors. If a molecule resembles a signal molecule in its steric and spatial properties, it may react with the cognate receptor in such a way as to "jam" its active

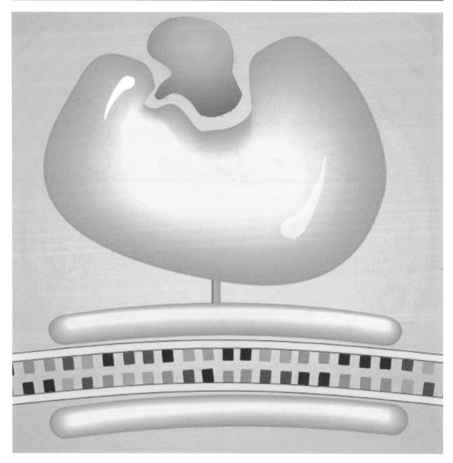

Fig. 34 Conceptual drawing of a receptor protein that accommodates a signal molecule in its active site and undergoes a conformational change that allows it to transcribe specific genes while the signal remains in place

site and preclude transcription of the genes that would normally be expressed (Fig. 35, bottom). Many biofilm inhibitors have been described in the natural world (de Nys et al. 1995), and we tend to emphasize those that completely inhibit biofilm formation, but we should realize that the inhibition of gene expression can be equally as effective as its activation in the control of subtle community processes.

The elegant work of Barbara Iglewski and Peter Greenberg and their small armies of acolytes has shown that Gram-negative bacteria produce AHL signals that control many processes, including biofilm formation, in neighboring cells of the same species (de Kievit et al. 2001). The details of these interactions are less important than the principle that individual cells in microbial communities send and receive chemical signals that affect the metabolism

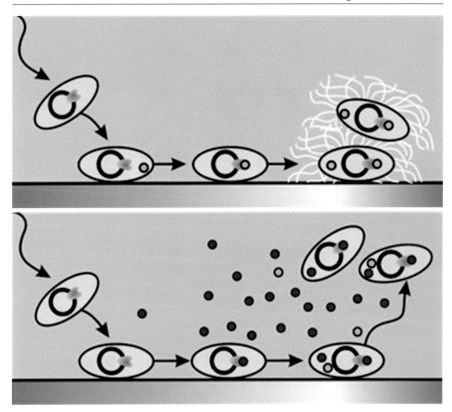

Fig. 35 Conceptual drawing describing the role of signals and signal inhibitors in biofilm development. *Top*: in natural environments a planktonic cells settles on a surface and produces a signal (*green*) that reacts with its cognate receptor and initiates the expression of genes that enable biofilm formation. *Bottom*: in the presence of high concentrations of a specific inhibitor (*blue*), the signal (*green*) is unable to react with the receptor, and the cells are "locked" in the planktonic state

and behavior of cells of the same species of different species and (probably) of eukaryotic hosts. In the simplest case, small soluble signal molecules would be released by a bacterial cell and diffuse to neighboring cells in the radial pattern dictated by the implacable laws of physics. Dazzo et al. (Gantner et al. 2006) have shown that cells of the same species can send and receive AHL signals over distances of $> 70\,\mu$m along the planar surface of a plant root, so we conclude that biofilm cells within 4 to $6\,\mu$m of each other must literally bombard each other with chemical chatter. We are therefore now justified in taking this conceptual model of chemical gradients of signal molecules produced by single cells, or by coordinated clonal groups of cells, and studying the behavior of cells of the same and other species that find themselves in these gradients. This concept can help us understand how primary colonizers can recruit mobile secondary colonizers by providing the gradients within which the mobile cells can perform their graceful arabesques of tumbles and runs and eventually find connubial bliss in a metabolically integrated community. Besides explaining how the ballet dancers and billionaires of the microbial world find each other, this concept of an infinitely complex overlapping system of signal gradients provides a conceptual basis for metabolic integration, and even for programmed detachment.

A chemical analysis of signal inhibitors that affect biofilm formation offers a tantalizing glimpse at dozens of chemical structures that bear no resemblance to known signal molecules and suggests that we may find many new classes of signals by finding their inhibitors first! We surmise that only in situ studies using probes that detect the signals themselves, or reporter systems that detect the expression of signal-producing and signal-receiving genes, will allow us to unravel the system that controls community development in prokaryotes. This task will be more complex than that addressed by eukaryotic embryologists because the process is guided by all genomes of the species that comprise any prokaryotic community, and we begin to think of niches occupied by giant shadowy metagenomes. The analogy of signals to hormones, in the biofilm context, implies a much higher level of organization in biofilm communities than has heretofore been contemplated. Taken to its logical end point, this concept would presume that the partners in a particular biofilm would have the genomic keys to produce mixed-species microcolonies, to build a metabolically integrated consortium, and to locate their operation in a favorable location vis-à-vis water channels. Decisions on detachment, or on swarming excursions, would have to be based on signal communication between sessile cells of several species, and this hypothetical cooperation could only work if all partners had the appropriate signal synthesis and signal reception machinery. If this degree of fantasy seems inappropriate, it may be useful to try to imagine human endocrinologists and immunologists, in the 1960s, sitting down to write about the coordinated processes of embryology or the subtleties of the inflammatory response before most hormones or any cytokines had been discovered.

If the virginal tissues of a newborn mammal release a gradient of a bacterial nutrient, like urea or glucose, they will trigger a stampede of hundreds of bacterial species. If they release signals, like those released to attract Rhizobium species to the roots of nitrogen starved plants (Long 2001), preprogrammed bacterial species will be attracted preferentially and rapidly and we will find specific populations on specific tissues. We find predominant populations of lactobacilli on the vaginal epithelium, and of *Staphylococcus epidermidis* in the skin of very young humans, and we now find that each mammalian species seems to attract and harbor its own particular species of staphylococcus. In environmental ecosystems, primary colonizers use signal gradients to recruit metabolically cooperative partners into spatially organized communities (Fig. 1) with unparalleled levels of physiological efficiency. Microbial communities have the unique property of being able to dissociate in a niche that has become unattractive and to reassemble in a new and more favorable downstream location, and this very useful ability can readily be explained in terms of signal gradients. The detachment signals of some or all of the component species would be produced at a constant level, and would be removed from the community by simple leaching, as long as the bulk fluid flow that delivers nutrients to the community remained at high levels. When bulk fluid flow diminished, and stagnation produced a lack of nutrients and an accumulation of waste products, the detachment signal would reach a critical level and trigger the conversion of the sessile cells of some or all species to the mobile planktonic phenotype. This involvement of a signal gradient in the complex and dynamic architecture of microbial communities leads to the hollowing of microcolonies in single-species biofilms (Fig. 26 and www.springer.com/978-3-540-68021-5: Movie 10) because the detachment signal tends to accumulate in the centers of these aggregates. The "seething" behavior of recently liberated planktonic cells can be seen in Movie 11 (www.springer.com/978-3-540-68021-5), and the frequency of the "hollowing" of individual microcolonies seen in the same material attests to the frequency of these detachment events. The simple cessation of flow is often sufficient to trigger massive detachment of cells from single-species *P. aeruginosa* biofilms, and we suggest that major sloughing events in macroscopic natural biofilms may be caused by similar signal gradients. The presence of large numbers of planktonic cells of different bacterial species, at specific times of the day, also suggests that the massive sessile communities that occupy surfaces in all rivers may undergo detachment in complex diurnal patterns.

2.2
Targeted Signaling in Microbial Biofilm Communities

While diffusion is sufficient to explain assembly, coordinated function, and detachment in the simplest microbial communities, this mechanism seems insufficient to explain the coordinated swarming behavior and complex fruiting body formation of the myxobacteria. These organisms gallop through the microbial jungle of soils and other complex environments in coordinated squadrons that move quickly, exclude interlopers from resident species, and react to nutrient conditions as an integrated community. In our recent examinations of these dynamic organisms, we have discovered that they produce membrane-bound vesicles that form virtual "bubble trains" between cells (Fig. 36, top), and electron tomography has shown that some of these "chains" constitute de facto tunnels between cells (Fig. 36, bottom). As we resolve ever more details of these chains of vesicles, it appears that they may be aligned along piluslike filaments (Fig. 36, top), which remind us of the microtubules and microfilaments that align and propel organelles in cyto-

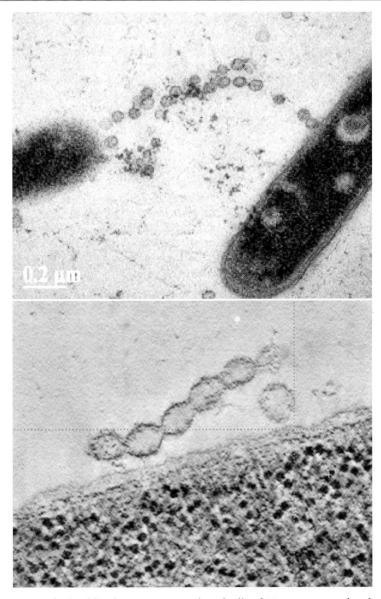

Fig. 36 TEMS, obtained by electron tomography, of cells of *Myxococcus xanthus* frozen at high pressure to minimize fixation artifacts. The tomographs are produced by integrating images obtained at different angles, to minimize steric aberrations, and they show well-defined chains of vesicles similar to those seen by Mashburn and Whitely (2005) to be involved in signal transfer in *P. aeruginosa*. *Top*: a chain of vesicles extends from the cell at the *bottom right* to another cell at *9 o'clock*, and the chain appears to be associated with a piluslike structure. *Bottom*: a similar chain of vesicles is seen to constitute a de facto tube, and the terminal vesicle (*left*) is seen to contain an organized electron dense structure. (Courtesy Manfred Auer and Jonathan Remis)

plasmic streaming in plant and animal cells. Mashburn and Whiteley (2005) have shown that similar vesicles carry quinolone signals between cells in biofilms formed by *P. aeruginosa*, and we suggest that this mechanism of cell–cell communication may operate in all bacterial communities but may reach its epitome in dynamic myxobacterial communities. The advantage of this method of communication, over simple signal diffusion, is that the signal strength would not diminish with distance and that the vesicles could be targeted to deliver their contents to particular cells and not to others. If we can stretch our imaginations to the levels required by these observations, we can imagine a mixed-species bacterial community in which cells of one species could produce vesicles that would deliver signals, at full strength, only to cells of the same species throughout the community. Simple "same-species" targeted vesicles could even be deposited in specific locations, like addressed letters, and could trigger specific reactions only in cells of the target species when they reached that location.

The observation that cells in myxobacterial biofilms form chains of vesicles between cells (Fig. 36) and that similar vesicles carry signals between cells in Pseudomonas biofilms raises the possibility of functional integration in microbial communities to a new level. It is now possible to visualize mixed-species communities (Fig. 11) in which cells can recruit new species into the community by releasing pheromonelike signals that induce positive chemotaxis, followed by biofilm formation and structural integration. Within structurally integrated communities functional integration could be achieved by the ability of each cell to produce a vesicle full of any signal and to target it to trigger any signal-controlled activity in cells of the same species, or of any other species, or of the host. This advanced form of communication could modulate any physiological property of the receiving cell, so that the production and activity of any enzyme necessary for mutual metabolic success could be controlled by individual community members. The sophistication of targeted signaling approaches that seen in complex multicellular eukaryotes and profoundly changes the position of prokaryotic organisms in the overall scheme of living things.

2.3
Other Signaling Mechanisms in Microbial Biofilm Communities

The discovery (Gorby et al. 2006) of the nanowires that transmit electrical energy from energy-rich to energy-poor regions of *Schewanella oneidensis* biofilms adds a particularly intriguing piece to the jigsaw puzzle of communication in microbial communities. Besides solving the long-standing problem of how thick biofilms can thrive on surfaces that supply insoluble nutrients only to those of its members who can make direct contact, this discovery places very long ($> 100\,\mu m$) conductive structures in microbial communities. If these structures are confined to the genus *Schewanella*, one can easily imagine how the transmission of electrical energy could, in itself, constitute a signal to cells of this genus in a mixed-species biofilm. *Schewanella* cells that found themselves in direct contact with the iron salts that they are so exquisitely equipped to reduce could send an electrical message to cells of the same species elsewhere in the mixed-species community that would report their happy circumstances. If these remarkable structures are not confined to that genus, then we can equally easily visualize a tangle of nanowires that would traverse the community and effectively connect cells of the same or of related species. The transmission of simple electrical pulses has been demonstrated, by the Lovely and Gorby teams, but it would only take one more effort from our exhausted imaginations to speculate that the same physical structures could carry modulated electrical pulses (www.springer.com/978-3-540-68021-5: Movie 3) that would constitute sophisticated electrical signaling.

2.4
Commensal Integration with Eukaryotes

It is axiomatic in microbial ecology, and in all ecological sciences, that interactions between living creatures often determine the success or failure of organisms in natural ecosystems. Pathogenicity and parasitism define relationships that are harmful to at least one of the organisms, and commensalisms and symbiosis define positive relationships in which one or both partners benefit. While medical and dental microbiology have, for good historical reasons, concentrated on negative relationships, microbial ecologists have studied positive relationships between microbes and higher organisms very extensively. We can choose from dozens of examples, but the symbioses between Rhizobia and nitrogen-fixing plants (Long 2001), and between "captive" cellulose-digesting bacteria and their insect (Bresnak and Brune 1994) and mammalian hosts, are among the most instructive.

In the bacteria/plant nitrogen-fixation system we have the best understood cooperative strategy that is hard-wired into the genomes of both partners. In conditions of nitrogen sufficiency, the plant roots and the nitrogen-fixing bacteria coexist, but they do not interact. When the plant senses nitrogen deficiency, it emits a signal that attracts only its partner species among the many Rhizobia in that particular soil and induces planktonic cells of that species to form a biofilm at specific locations on its root surface (Gantner et al. 2006). The presence of the sessile cells of the bacterium then induces anatomical changes that lead to the formation of a very elaborate "infection thread" that penetrates the root to initiate nodule formation. There is no benefit to either the plant or the bacteria in the very complex "dance of the seven veils" that precedes nodule formation and the start of nitrogen fixation, but the genomes of each partner are programmed to interact for eventual mutual benefit. This example teaches us that signal synthesis may be under environmental control, that the signal-receiving partner may be selected at the species level, and that the interaction may be iterative and almost as complex as embryogenesis, long before mutual benefits are realized.

Animals that digest cellulose do not do so using their own enzymes; they recruit cellulolytic bacteria and provide an optimal environment for their diminutive partners to digest plant material, or even wood, for their mutual benefit. In the cow, the bacterial processing of plant material takes place in a four-chambered stomach whose first chamber teems with cellulolytic bacteria that instantly colonize cellulose and form biofilms that focus their enzymatic attack with awesome efficiency (Fig. 25). This well-defined captive ecosystem (Cheng and Costerton 1981) is fastidiously anaerobic and deficient in nitrogen, which it requires in the form of ammonia. So the host must accommodate a fastidious anaerobic process within its highly aerobic tissues, which it must supply with oxygenated arterial blood, and it must also supply its small but essential guests with sufficient nitrogen in the form of ammonia. K.J. Cheng and I discovered that ruminants "employ" another bacterial population (Fig. 37) to solve these problems and that a special biofilm lines the rumen, scavenges oxygen that would kill the cellulolytic population, changes urea in the blood and tissues to ammonia, and (for good measure) feeds itself from the sloughing cells of this squamous epithelium! We found that this very well-defined four-species rumen-lining biofilm population establishes itself within the first 4 d of life, from a maternal source, and that gnotobiotic (germ-free) animals that have no bacteria in their gut fail to develop rumen structures and die of urea poisoning. We have not discovered the signal(s) that mediates colonization by this facultatively aerobic ureolytic biofilm population, but we have established that a highly specialized cellulolytic ecosystem can be accommodated in a mammalian body with the assistance of another symbiotic bacterial population. The cellulolytic bacterial population that occupies the hindgut of the termite, and allows this creature to eat your house, is even more fascinating (Bresnak and Brune 1994) because it is more closely integrated with specific host tissues and processes cellulose in a sequential, as opposed to a mixed, system.

I clearly forgot my ecological training when I assumed that rat skin would be colonized by *Staphylococcus epidermidis* and used this human commensal organism as a control in animal experiments concerning the pathogenic col-

Fig. 37 Conceptual drawing of the facultative commensal bacterial population of the rumen wall and of the strictly anaerobic cellulolytic bacterial population of rumen. The wall population protects the anaerobic population from oxygen from the blood and supplies essential ammonia by the reduction of circulating urea, which is toxic to rumen organisms. The wall population sustains itself by digesting sloughed cells from the rumen epithelium, is acquired from the mother in the first few days of life, and stimulates the full anatomic development of the four-chambered stomach in ruminants

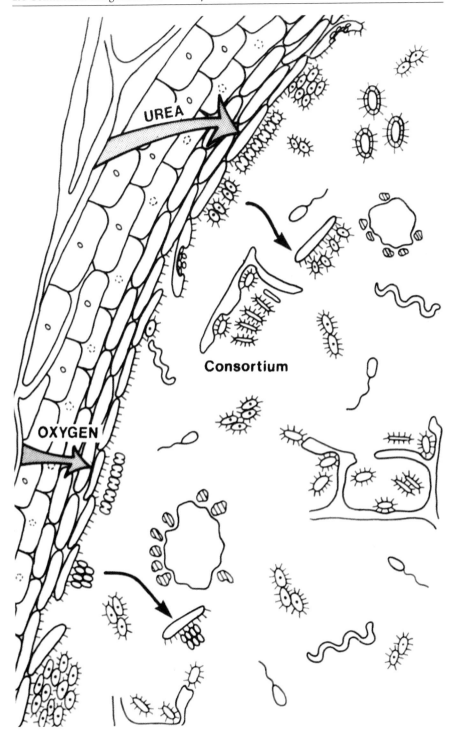

onization of vascular catheters. Working with John Olerud's group, we have found that rat skin is colonized by its own staphylococcal partner (*S. xylosus*), and we anticipate that we will find that each mammalian species has recruited and tamed its own particular bacterial skin symbiont. When we find a stable "climax" population (ecological term) on the surface of a plant or animal tissue, we should invoke the rich history of positive eukaryote/prokaryote cooperations and search for the signals that mediate this commensal relationship very early in the life of the eukaryotic partner. The spatial relationship of the commensal bacteria to the cells of the host tissue may be very intimate, as in the case of *S. epidermidis* in the human skin (Figs. 38 and 39) and the six species of lactobacillus (*L. rhamnosus*, *L. casei*, *L. jensenii*, *L. plantarum*, *L. fermentum*, and *L. acidophilus*) on the human vaginal (Zhong et al. 1998) epithelium (Fig. 40), and the interesting question concerns the immune privilege accorded these minute guests. Clearly, the antigens of commensal bacteria would be presented to the host immune system early enough for them to be categorized as "self", and this may explain why the subcutaneous tract surrounding a Tenckhoff catheter may be colonized by billions of cells of *S. epidermidis* without the slightest signs of inflammation. Similarly, the vaginal epithelium may entertain billions of lactobacilli without any inflammatory reaction but blow up like a balloon when interlopers of an unfamiliar species make incursions. This is logical, but the unanswered question is how commensal bacteria are protected from the innate defenses of the host (antibacterial peptides and surfactants), and it may be a sine qua non of the commensal role that the chosen species be resistant to these factors. One of the central themes of this book is that strategies that work are usually repeated throughout the microbial world, and we should look for signal-directed colonization by selected bacterial species instead of prattling on about glucose-loving bacteria being attracted to glucose-producing tissues.

The maintenance of the structural association of a nonmotile bacterium (*S. epidermidis*) with the human epidermis, which sloughs epidermal cells at a relatively high rate while replacing them from the stratum corneum, has been the subject of conjecture since the association of biofilm bacteria with the skin was first discovered. The retention of these bacteria in this dynamic tissue environment may be partially explained by our very recent discovery that cells of *S. epidermidis* make remarkable proteinaceous honeycomb structures that resemble host tissues in complexity and scale (Figs. 14–17) (Sect. 1.2.1). If, as we suspect, cells of *S. epidermidis* make these very extensive tertiary structures and integrate them with the similar collagen and elastin structures of the skin (18), the bacteria would not be trapped by sloughing skin cells but would have access to a structured intercellular highway system that ramifies between the epithelial cells and may even extend deep into the dermis. We are presently comparing the genomes and expressomes of honeycomb-forming strains with those of strains that have lost the

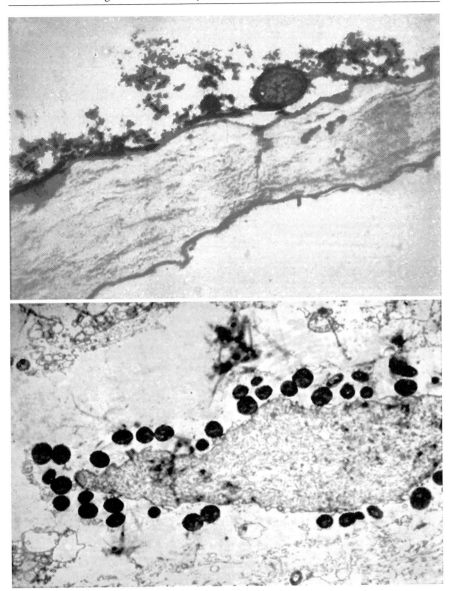

Fig. 38 TEMs of human skin showing the integration of biofilms formed by *S. epidermidis*. *Top*: TEM of dry skin (forearm) from Kan Lam, showing the presence of a sessile Gram-positive bacterial cell adherent to a superficial skin cell by means of condensed matrix material. *Bottom*: TEM of an extensive biofilm of Gram-positive bacteria on a skin cell deep ($\pm 70\,\mu$m) in a moist area between Bill Costerton's toes. Do not attempt this at home

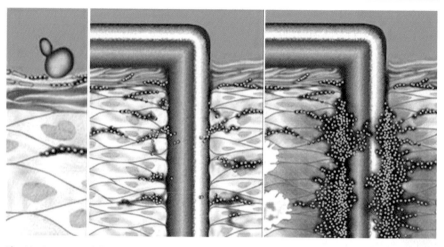

Fig. 39 Conceptual drawing of microbial colonization of human skin. In the *left panel* cells of *S. epidermidis* (*black*) are seen to inhabit the deeper layers of the skin, while cells of this species and of Gram-negative bacteria and fungi (*blue*) all occupy the distal layers of this squamous epithelium. The *central panel* shows that, when the skin has been prepared for surgery and a staple has been inserted, the surface of the skin is uncolonized, but living biofilms of *S. epidermidis* occupy the deeper layers in the vicinity of this foreign body. The *right panel* shows the development of an extensive *S. epidermidis* biofilm on the surfaces of the staple and the initiation of a mild inflammatory response involving the mobilization of leukocytes

ability to make these structures, in an attempt to identify the unique bacterial genes that control the architecture of these remarkable structures. If, as is very likely, the ability to make tertiary structures is a property of many (most?) bacterial species, then bacteria may not always be the passive "tenants" in mixed-kingdom communities. We must reexamine the structures of sponges and lichens, using markers for prokaryotic cytoplasm, to determine whether bacteria may build some structural components of these communal organisms, or even whether prokaryotic cells can build structures that accommodate eukaryotic partners.

If one asks the painful question of why this basic ecological perception has been so long in coming to the notice of medical microbiologists (including myself), the answer involves the timetable of recent discoveries in biofilm microbiology. Since the mid-1880s we have studied bacteria in single-species cultures, while recording the sources from which they were isolated, but the ecological concept of commensalism was not well developed in our community. In 1978 we discovered that bacteria live predominantly in matrix-enclosed biofilm communities on surfaces (Costerton et al. 1978), including the surfaces of mammalian tissues. In the 1980s we discovered that virtually all bacterial species can communicate with each other by means of chemical signals (Fuqua and Greenberg 2002), and in 1998 we discovered that many of

Fig. 40 Light micrograph of a "raft" of epithelial cells recovered from the vaginal epithelium of a healthy human volunteer. Fluoresin staining reveals the presence of a biofilm composed of large numbers of rod-shaped bacterial cells, embedded in an extensive EPS matrix, that contains "tower" structures at least 35 μm in height. This natural commensal biofilm virtually occludes the surface of the host tissue, whose location can be deduced by the buried orange-colored nuclei of its component cells

these signals controlled biofilm formation (Davies et al. 1998). The stage was then nicely set for the concept that a plant or animal could secrete a chemical signal that mobilizes its prokaryotic partner from a multispecies population and directs this bacterial species to form a biofilm on a specific surface and then to initiate its role in an interkingdom partnership. Timing and communication would seem to be of the essence.

The success of the mammalian female reproductive system is attested by our presence on Earth, and it is interesting to ponder the whole evolutionary process that led to its evolutionary development. The reproductive strategy of producing and extruding very large numbers of eggs, with cunning designs to favor fertilization by sperm over infection by bacteria, needed to be modified for the overall emergence of child-rearing mammals and humans. Mammalian reproduction is a triumph of the harnessing of microbial ecology because sperm are effectively sterilized by the acids and surfactants produced by the specialized resident bacterial population of the vagina and the egg is ushered to its rendezvous on the endometrium without effective bacterial challenge. Medical microbiology has trivialized this triumph of effective microbial ecology by declaring the uterus to be sterile and by invoking a mysterious "cervical mucus plug" that somehow excluded bacteria but permitted the passage of sperm through the gateway of the cervical os. How this plug functioned in the hydraulic furor of actual sexual intercourse is best left to the desiccated imaginations of the nice laboratory-based people who offered this fatuous explanation to generations of ecologically gullible, but otherwise intelligent, medical students!

While the tacit acceptance of a stupid explanation may not seem like a cardinal sin, the opportunities lost when we accept an idiotic concept and fail to examine the real explanation(s) of a natural phenomenon may rate a hearty mea culpa. When we noted that amphibian skin resisted bacterial invasion, even though frogs and toads live in the most challenging of environments, we discovered the first of the antibacterial peptides that are now known to comprise a bulwark of our own innate defenses. When we noted that biofilm-coated contact lenses were clean and sterile minutes after insertion into the eye, we discovered that these "amphibian" molecules also defended the human conjunctiva. Now that the sterile cervical plug has gone the way of Grimm's fairy tales, and we know that the distal parts of the human uterus is colonized by bacteria, we could learn a lot by enquiring why Zell McGee's Gonococci can invade and scar the fallopian tubes (McGee et al. 1999) while lesser organisms cannot. We know that the prostate glands of men in the age cohort of 55 to 60 years are all colonized by a rich bacterial flora, growing in biofilms in the ascinar spaces, and we know that the prostate gland is adjacent to the bladder but that cystitis is rare in men of this (or any) age. The simple mapping of the bacterial population of the male urinary system would make sense, using the same methods (Wagner et al. 2003) that Mickey Wagner and his ecological colleagues have used to map their cave stream, and we could examine the mechanism that stops bacteria from making the 2-mm voyage from the prostate to the bladder. My personal experience with aging convinces me that it is not hydraulic force that prevents bacterial access to the male bladder, and the real answer may involve a discovery as profound as some new component of the innate defense system. The crucial element in ecological research is the simple mapping of the bacterial component of any given ecosystem, and Joe Sung's comprehensive mapping of the bacterial populations of the biliary system of the cat (Sung et al. 1991) yielded basic and surprising information. This ecological approach also allowed Joe to challenge the feline biliary system with extraneous bacteria and to explain the etiology of bacterial liver diseases in humans. Joe is presently chairman of the Department of Medicine in the Prince of Wales Hospital in Hong Kong.

When we began to examine bacteria in the communities in which they actually live – in natural systems – we lifted the lid off of microbiological concepts that had served us for more than 160 years. Evidence was available for the predominance of biofilms (e.g., the dental literature), for cell–cell signaling (e.g., *Vibrio harveyii*), and for social behavior and the formation of complex structures (e.g., the myxobacteria), but we still conceived of most bacteria in terms of the behavior of planktonic cells in liquid culture. These concepts failed to explain the operations of natural ecosystems or the etiology of chronic infections, but our main tools were predicated on the recovery of bacteria from nature and their cultivation in single-species cultures. At this moment, microbiology is the most challenging of modern sciences because our conceptual core needs rebuilding and we have the tools to extend our concepts of what bacteria can accomplish simply by generalizing from firm new data obtained from one or more organisms. Biofilms predominate in natural and pathogenic systems (Parsek and Singh 2003), and antibiofilm strategies will soon prevent and cure chronic biofilm infections. Cell–cell signaling controls most bacterial activities, including biofilm formation and the formation of functional communities that may combine many prokaryotic and eukaryotic species, and we can already control community behavior by signal manipulation. We can identify species, assess viability, quantify metabolic activity, and even determine interspecies and interkingdom interactions by direct observation and without dependence on isolation or culture methods. Bacteria are capable of producing very complex structures, on a scale much larger and more complex than their own cells and biofilms, and they may use these structures in the construction of communities that include natural and transformed eukaryotic cells. We really don't know what bacteria are capable of, having been limited in our concepts until very recently, but we know that their upper limits have now been shifted massively, to levels previously considered to be the preserve of eukaryotic organisms.

3 The Microbiology of the Healthy Human Body

The normal tissues and organ systems of the healthy human body have all been examined using culture methods for bacterial detection and recovery, and we will now examine the colonization patterns that they have deduced in the light of data derived from ecological concepts using modern methods of direct observation and molecular biology.

3.1
The Human Integument

Conventional medical microbiology views the human skin as a basically sterile squamous tissue whose shedding surface is colonized by a motley collection of bacterial and fungal species that can feast on the relatively refractory keratin and oil banquet in a relatively dry environment. Skin that has been prepared for surgery yields reassuringly sterile swabs because individual bacterial and fungal cells that would grow in culture have been killed in situ or by residual sterilants picked up on the swab, and the surgeon's scalpel is seen as slicing through sterile tissue. But direct examination of human skin has revealed (Fig. 38) cells of *Staphylococcus epidermidis* growing in biofilms between the squamous cells of the skin, at least five skin cells deep in dry areas and 15 to 20 skin cells deep in moist areas such as those surrounding surgical stoma. The surface of rabbit skin yields negative cultures on swabbing, after it has been prepared for surgery using iodine or Hibitane, but rinsed and mascerated skin yields ca. 1×10^5 cells/cm^2 of *S. nepalensis* on culture, and very large numbers of this organism are seen on direct examination of "prepped" skin. We conclude that human skin contains a structurally integrated population of cells of *S. epidermidis* and an adventitious population of other bacteria and fungi (Fig. 39), and that sterile preparation kills the adventitious population without affecting the integrated population. Therefore, when a medical device is placed across prepared skin, the living bacteria in the integrated population will inevitably form biofilms on its surface (Fig. 39). John Olerud's group at the University of Washington has joined us in the direct examination of rat skin, and we find large numbers of cells of *S. xylosus* and *S. lentus* in the surface layers of this tissue, and we occasionally see areas in which these cells have formed (Fig. 18) amorphous structures reminiscent of the honeycomb structures formed (in vitro) by the lymphoma-associated strain of *S. epidermidis.*

The picture that emerges is one that is very familiar to microbial ecologists working with plants and animals in their natural ecosystems, in that each mammalian species appears to have arrived at a commensal arrangement with its own "tame" staphylococcal species. The bacterial partner in the commensal arrangement benefits from continued exclusive access to an attractive ecological niche that continues for many (human) generations, and the host mammal benefits from an entrenched bacterial partner that competes (usually very successfully) with potential pathogens. It is entertaining to speculate concerning the conversations that *S. epidermidis* has with other bacteria, in the skin glands and hair follicles of human teenagers with raging hormonal changes, concerning which bacterial species will dominate and facilitate flawless beauty or cratered misery. Commensal bacterial partners may be recognized as "self" by their hosts, but *S. epidermidis* is only cooperative in skin and is as pathogenic as any other staphylococcal species when it penetrates other tissues. It will be very interesting to discover whether the close association of each mammalian species with its own commensal strain of staphylococcus is the result of mutual adaptation or whether there is a signaling mechanism that facilitates this association very early in life.

Teleological and ecological thinking leads us swiftly to the conclusion that the human eye is a truly remarkable organ whose interface with the integument is critically important. The survival value of the functioning eye is self evident and is attested by the convergent evolution that has produced very similar eyes in animal families as distant as the human (mammal) and the octopus (mollusk). The huge genetic investment of the eye would be completely negated if this organ failed due to bacterial infection of the conjunctiva, which protects the eye and functions as part of the integument but is challenged in Afghan tribesman and octopus alike by flying sand. Some very effective system for the control of bacterial colonization must have evolved, very specifically, to protect the human conjunctiva from bacterial colonization, and we see the modern manifestation of this mechanism in people who wear contact lenses. We have examined contact lenses, straight from 3-year-old solutions in storage cases in the fetid bedrooms of itinerant rugby players and their makeup-encrusted (occasional) partners, and we have found bacterial biofilms that virtually occlude the lenses. When these lenses are inserted into the bloodshot eyes, using the "sterile" tips of spit-moistened fingers, they can be removed after 20 min and found to be sparkling clean and utterly devoid of bacterial biofilms. Obviously, the human body protects its most vital organs with its innate immunity, surfactant chemistry, and enzyme cleaning powers, and the efficacy of this concentration of biofilm control factors in the human conjunctiva demands further study. Itinerant rugby players are numerous, unfailingly cooperative, and totally uncomplicated by independent thought processes!

3.2
The Human Female Reproductive System

Many human organ systems connect the core of the human body, which must be maintained in a healthy sterile condition, with the microbial free-for-all of the external environment. This concept presumes the existence of a colonization boundary, somewhere in the continuum of the organ system, where bacterial entry is controlled by host factors. The human female reproductive system is a case in point because the fallopian tubes react to the presence of bacteria by scaring that leads to infertility (McGee et al. 1999), and the ovaries lie (effectively) in the peritoneal cavity, whose sterility is sacrosanct. The female reproductive tract bears [sic] the additional burdens of accommodating to the intrusions inherent in sexual activity and to the maintenance of the resultant fetus in microbiological conditions that allow its survival. In conditions much less optimal than our modern maternity facilities, early human females conceived and bore children in a process that, taken at its most basic level, was designed to allow access of the sperm to the egg while protecting this precious cell from the voracious bacteria in its environment.

When we first combined our direct microscopic examinations (Fig. 40) with Tony Chow's cultural studies of the vaginal flora of 20 volunteers, it was clear that the phenomenal microbial population of this organ is dominated by the large square-ended cells of the lactobacilli. We later discovered that minor populations of *Staphylococcus aureus* (Fig. 41) and of *Gardnerella vaginalis* shared this ecosystem with the lactobacilli, but sampling at multiple sites and at various times during menstrual cycles always confirmed the predominance of these acid-producing commensal organisms (Sadhu et al. 1989). The native lactobacilli of the human vagina are *L. rhamnosus* and *L. ferementum*, and not the lactobacilli readily available in yogurt, which explains the greater efficacy of Gregor Reid's probiotic in the treatment of vaginosis caused by ecological upsets in which the resident lactobacilli are challenged by interlopers. The basic and ubiquitous venereal disease of humans is probably the ecological change that occurs when a woman's natural flora is challenged by that of her sexual partner's other partners, and an ecological study of this process is bound to be revealing and to promote fidelity.

The human uterus constitutes the effective boundary between the heavily colonized distal elements of the organ system and its normally sterile proximal organs, and as such it represents a microbial ecosystem under considerable stress from host defenses. We have never been able to culture bacteria from uterine tissues, or from intrauterine devices, but direct observations

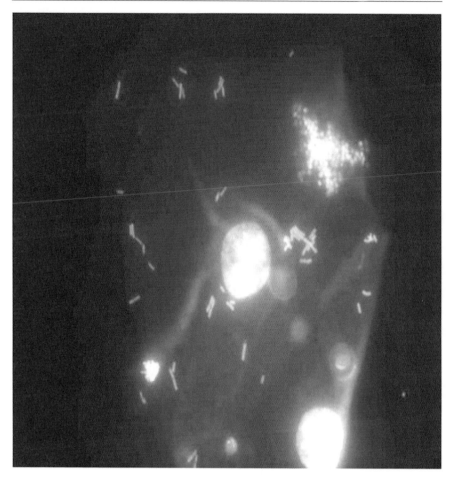

Fig. 41 Fluorescence micrograph of a single human vaginal epithelial cell showing the presence of both rod-shaped and coccoid cells that react with the "eubac" FISH probe that hybridizes with the 16 S rRNA of all eubacteria. Parallel studies, with specific FISH probes, showed that the coccoid cells in the microcolony were *S. aureus*

of the endometrial surface have shown the presence of very large numbers of biofilm bacteria, and direct observations of IUDs (Marrie and Costerton 1983) have revealed the thickest and most luxuriant biofilm we have ever seen on any medical device (Fig. 42). Many of the bacteria in the endometrial

Fig. 42 SEMs of biofilm that forms on intrauterine contraceptive devices (IUDs) recov- ▶ ered from human uteri. *Top*: detail of cellular structure of very thick (±2 mm) biofilms that contain bacteria of many different morphotypes, including that of Actinomyces. *Bottom*: expanded field showing presence of biofilm on both plastic and metal (copper) wire components of the device

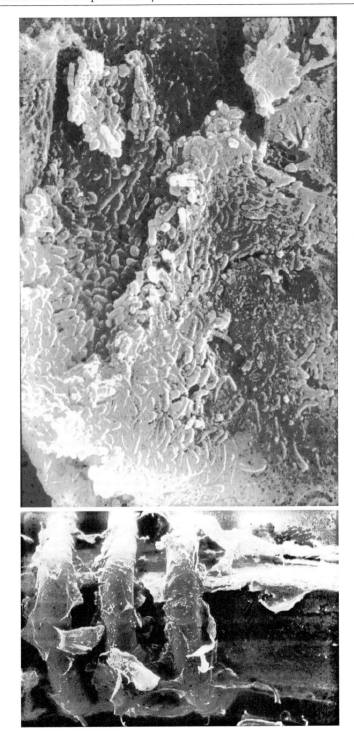

biofilm, which extends intermittently from the cervical os to the fundus, appear to lack normal cell walls, and we suggest that this adaptation may protect them from the neutrophils that constitute one of the major host defenses that limit further upstream colonization. The inert surfaces of IUDs extend into the proximal parts of the uterus and are connected to the vagina by a retrieval string, and they accrete biofilms exceeding 1 mm in thickness. Many IUD biofilms harbor large microcolonies of Actinomyces known to cause scaring of the fallopian tubes, and the millions of women who wore these devices for many years without tubal complications attest to the efficacy of the host defenses that confine bacterial colonization to the uterus. The colonization boundary in the human female reproductive system lies somewhere in the uterus, and direct observations and molecular methods will help us to locate that boundary and to protect human reproductive health.

We examined the effects of biofilm formation on the surfaces of IUDs, in connection with the Dalcon Shield litigation, and we placed sterile IUDs (surgically) in one of the bicornate uteri of rabbits while introducing matching devices through the other cervix in the same animal. When the rabbits were fertilized with sterilized sperm introduced directly into the uteri, the IUDs lacking bacterial biofilms were not contraceptive, and whole litters of pups were found right beside the sterile devices. We concluded that IUDs actually carry out their contraceptive function by accreting biofilms and inducing inflammation in the adjacent endometrium and that it is this inflammation, and not the physical presence of the device, that prevents the implantation of the fertilized egg. Now that we and others have shown that the human uterus is not sterile, the intriguing question of how the fertilized egg and the developing fetus adjust to the presence of microbial cotenants comes quickly to the fore. The choroid components of the fetus interface with areas of the endometrium known to be colonized by bacterial biofilms before they become the placenta, and preterm delivery has been linked (Romero et al. 2004) to the presence of bacteria in the allantoic fluids. Patients with a tendency to preterm delivery have even been treated with antibiotics, without statistically significant evidence of improvement (Espinoza et al. 2006). Roberto Romero and his NIH team have joined us in the examination of the microbial ecology of human pregnancy, and we have identified macroscopic flocs of biofilm bacteria in the allantoic fluid of affected women that may even form very extensive "sludge" in extreme cases. Microbiological cultures from the allantoic fluid in normal successful pregnancies have also shown the presence of bacteria, and we are presently initiating a study of normal and preterm fetuses in which we will map and characterize bacteria using direct observation and FISH probes. Our working hypothesis is that bacteria are present in the uterus, that they are associated with fetal tissues during development in utero, and that deleterious effects occur only if the fetus reacts to their presence with an inappropriate inflammatory reaction. If we find bacteria to be associated with the fertilized egg in the window of time surrounding implantation, it will be necessary to rethink our contention that sperm must be sterilized by bacterial factors during their passage through the vagina.

3.3
The Human Urinary System

Lori Graham's PhD thesis examined, with the help of a dedicated group of volunteers who could contribute "clean catch" samples of only about 10 ml of morning urine, the biofilm colonization of the human female urinary tract. Samples from this intrepid band showed that one third of the human cells that had sloughed during the night were covered, on one side only, by a very well-developed biofilm composed predominantly of the large square-ended rods characteristic of lactobacilli (Fig. 43). These data allowed us to place the colonization boundary of the female system proximal to the distal one third of the urethra, but there is no convenient histological transition in this region to which we could attribute local colonization resistance. We concluded that this urethral barrier population corresponded closely with the vaginal population of the individual, and we note that this acid-forming community is usually effective in precluding the upstream colonization of what is (in fact) really a very short open tube from the exterior to the bladder. In instances where the anatomy compromises the ability of the bacterial barrier population to keep bladder colonization at bay, Katerina Eden and her colleagues have had some success in preventing pyelonephritis by irrigating the whole bladder and colonizing it with vaginal lactobacilli (Wullt et al. 1998). So the barrier population can be moved from the urethra to the bladder without loss of efficacy, and we have found that it is equally effective in protecting the kidneys by colonizing surgical ilial conduits (Chan et al. 1984) that carry urine to external stoma following removal of the bladder. The colonization barrier of the female urinary system has been located, but not explained, and it is interesting to note that "stone bruising" of this region of the urethra may lead to "honeymoon cystitis" when the tissues responsible for this activity are mechanically damaged. The male urethra is less vulnerable, simply because of greater length, but it is often compromised later in life by the tendency of the prostate gland to act as a refuge and reservoir for a wide variety of organisms.

The colonization boundary of the urethra is immediately compromised by the insertion of a Foley catheter, and it is significant that the risk of cystitis rises by ca. 10% per day for every day that this device remains in place. The biofilm barrier to upstream colonization can also be compromised by the use of broad-spectrum antibiotics, which kill commensal organisms as readily as they kill pathogens. The introduction of sterilants into the periurethral space outside the catheter retards the colonization of the bladder, but it is significant that similar improvements can be seen with simple better fitting of the catheter by optimal size selection (Khoury et al. 1989). Prevention of bladder colonization is not rocket science but is simply a matter of hydraulics and the maintenance of effective bacterial biofilm barriers. The upstream colonization of the kidneys, via the ureters, is mercifully rare and seems to require the mediation of "P" pili in *E. coli*, but pyelonephritis is a very serious inva-

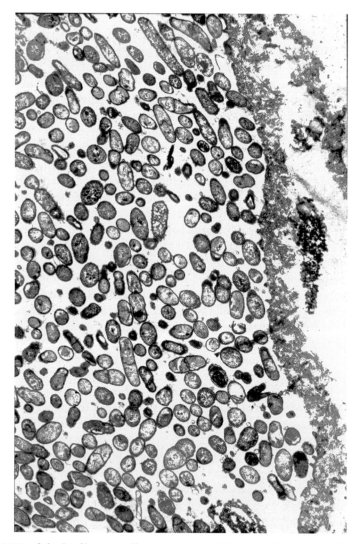

Fig. 43 TEM of the biofilm typically seen, on one side only, of epithelial cells sloughed from the human urethra during the night and recovered in morning "clean catch" urine. These very thick biofilms, whose component cells have Gram-positive cell walls consistent with those of lactobacilli, occupy the surfaces of one third of the human cells (*right*) in morning urine, and we conclude that the distal third of this organ is colonized by these commensal biofilms (Courtesy Lori Graham)

sion of a core organ that does not function well when colonized by bacteria. If the organism colonizing the hilus of the kidney is *Proteus vulgaris*, the natural tendency of this bacterium to produce struvite crystals causes the "petrification" of its biofilms (Fig. 44) to form "staghorn calculi" that virtually fill the central spaces of the kidney. Mortalities exceeding 50% are encountered

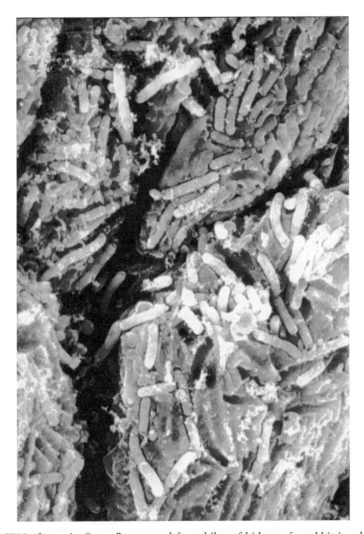

Fig. 44 SEM of struvite "stone" recovered from hilus of kidney of a rabbit in which experimental pyelonephritis had been induced with *Proteus vulgaris*. The biofilms produced by this organism become mineralized with magnesium ammonium phosphate (struvite), and the individual bacteria live in discreet "caves" within these stones, which are retained in the kidney in the form of a lethal "staghorn calculus". Other urinary and biliary stones have similar structures and etiologies (Courtesy Curt Nickel)

if petrified biofilms are allowed to form in the kidney, and the placement of "T" tubes to drain urine from infected kidneys has produced "struvite factories" that produce as many as three large struvite masses per week. So the human urinary organ system is a nonscenic waterway with a very effective biofilm barrier in its distal centimeter (three centimeters in the male) that precludes upstream bacterial colonization for many years in most healthy individuals. Mechanical damage to tissues, or ecological damage to bacterial populations, may destroy this barrier, but it can be reestablished and maintained with proper ecological management.

3.4
The Human Biliary System

The biliary system is virtually definitive of the barrier concept because it connects a very highly colonized organ (the intestine) with a core organ (the liver) that is rapidly and severely damaged by any bacterial incursions. Joe Sung described the barrier(s) to colonization in the human biliary system in his PhD thesis in my lab at the University of Calgary. First we examined the biofilms that physically block biliary stents that are inserted into the bile duct to keep it patent if it is compressed by developing pancreatic tumors, and we found very thick accretions in which classic multispecies biofilms (Fig. 45) alternated with layers of crystallized bile (Sung et al. 1993). These structures closely resemble those of the brown pigment stones (Leung et al. 1994) that develop spontaneously in the human biliary system, and their formation contradicts the old adage that the antibacterial properties of bile keep bacteria from colonizing this important organ system. Biofilms develop very rapidly and very luxuriantly, on inert surfaces, in the presence of the undiminished flow of "full-strength" bile.

Joe reasoned that the sphincter of odi, which prevents reflux of bile where the bile duct joins the ileum, might play a role in colonization resistance, so he removed this structure (Sung et al. 1992) and studied upstream bacterial migration in cats with and without this sphincter. When we placed an inert plastic surface in the gall bladder, to act as a "trap" to allow biofilm formation by any bacteria that had ascended the distal bile duct, we found that bacteria made excursions up the bile duct at least once every 2 weeks, in the presence or absence of the sphincter of odi. If there are no inert surfaces for the bacteria to colonize, they simply wash back out of the bile duct after challenging the Kupffer cells of the liver with numbers of bacterial cells that they are (obviously) capable of processing by phagocytosis. Joe then challenged the liver by the introduction of *E. coli* cells directly into the posterior vena cava (Sung et al. 1991) and found that this organ could withstand the incursion of 1×10^4 cells but that the introduction of 1×10^5 cells by this route caused severe cholangitis. We conclude that the barrier to upstream colonization of the

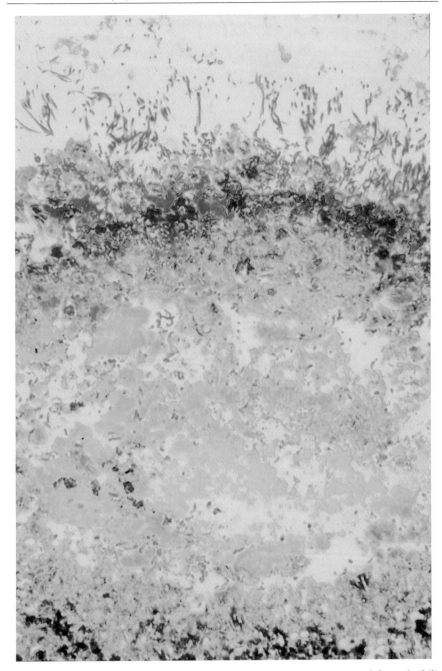

Fig. 45 Light micrograph of the material occluding a biliary stent removed from the bile duct of a pancreatic cancer patient in whom it had lost patency. This micrograph shows concentric layers of cell-rich biofilm (*top*) that alternate, throughout the mass, with layers in which large amounts of bile salts (*yellow*) have been deposited

human biliary system resides with the properties of the epithelia of the duct, and not with any biofilm barrier population. We note that bacterial infections of the liver are relatively rare, in the absence of instrumentation of the bile duct, and we conclude that this system must be very efficient.

3.5
The Human Pulmonary System

The microbial ecology of the pulmonary system is perhaps more complex than that of any other human organ system, and the apparent underlying strategy is to control the nature of the bacterial challenge to the lung rather than to maintain anything like sterility. Air enters via the oropharynx, whose epithelia normally bear a predominantly Gram-positive biofilm in healthy individuals, and this space communicates with the sinuses and the middle ear, both of which maintain similar epithelial populations. The pattern of bacterial colonization of the oropharynx appears to depend on the affinity of certain Gram-positive bacteria for the tissue-bound fibronectin that covers these epithelia in healthy individuals, and any planktonic bacteria or biofilm fragments that are aspirated deeper into the system are likely to be derived from this population. Buzz Johanson showed that the increased salivary protease levels produced by stress remove fibronectin from the oropharyngeal tissue surfaces and promote a recolonization by many Gram-negative species that are potentially pathogenic in the lung or in the middle ear and sinuses (Woods et al. 1981). As in the urinary tract, the instrumentation of the pulmonary system by the installation of an endotracheal tube bypasses the colonization resistance of the system and connects the heavily colonized distal organ (oropharynx) with a susceptible core organ (the lung), often with disastrous consequences (Adair et al. 2004).

We have shown that as many as 1×10^5 individual planktonic cells that have been aspirated into the lungs of four different mammalian species (Morck et al. 1990) are completely cleared, so that the homogenized lungs of sacrificed animals yield negative cultures 20 min after exposure. The multiple bifurcations of the pulmonary tree and the action of the mucocilliary "escalator" both work to prevent the deep aspiration of larger, heavier biofilm fragments and favor the deep penetration of single aerosolized cells. The defenses of the lung are predicated on the ability of the neutrophils of the terminal bronchi and of the alveoli to kill incoming planktonic cells by phagocytosis (Hoiby et al. 1995), especially in the presence of bactericidal and opsonizing antibodies (Fig. 46a). These defenses were very effective when humans lived in more natural environments, but more recent exposures to biofilm-laden air conditioning systems have allowed a ubiquitous littoral saprophyte (*Legionella pneumophila*) to emerge as a major human pulmonary pathogen. Failures of the pulmonary defense system occur when biofilm fragments,

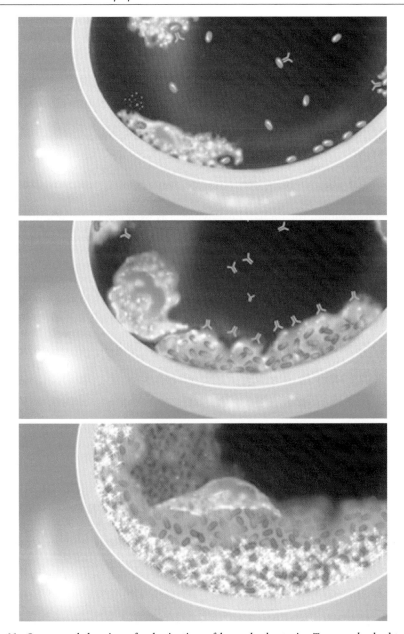

Fig. 46 Conceptual drawing of colonization of lungs by bacteria. *Top panel*: planktonic bacteria invade the alveolus and are killed by bactericidal antibodies (*yellow*) and by phagocytes following opsonization. *Middle panel*: when biofilm fragments are aspirated into the lungs, the bacterial cells are protected from killing and from phagocytosis, so that these sessile aggregates may persist for months or years. *Bottom panel*: proliferation of biofilm fragments to form extensive biofilms in the alveolus, and calcification often converts these sessile communities into permanent inclusions

from the oropharynx or from the environment, cannot be resolved by the normal phagocytic mechanisms (Jensen et al. 1990) of the deep lung (Fig. 46b) because of their size and their inherent resistance to this process. Biofilm cells that survive the phagocytic and chemical (Singh et al. 2002) "reception committee" of the deep lung may initiate an acute planktonic attack on the tissues (e.g., Pontiac fever), or they may assume a more cryptic strategy and calcify their microcolonies (Fig. 46c) to await the opportunity for a frontal attack when the host is weakened. The meloidosis infections caused by *Pseudomonas (Burkholderia) pseudomallei* are a case in point, in that victims are exposed to biofilm fragments containing sessile cells of this organism during rice cultivation, and microcolonies persist in the lung for very long periods of time (Vorachit et al. 1995b). We have shown that planktonic cells of this organism can overwhelm the human host, when it is weakened by starvation or by age-related deterioration of host defenses (Vorachit et al. 1995a), and the resultant acute pneumonia can be rapidly fatal. The pulmonary organ system is well adapted to handle almost continuous exposure to moderate numbers of individual bacteria evenly suspended in air, and the barrier population in the oropharynx can be seen as preventing local proliferation and biofilm formation by potential pathogens. Biofilm fragments, from the distal system or from the environment, constitute an invidious threat because they cannot be resolved by the phagocyte-based defenses of the lung and can emerge from containment and senescence to cause acute pulmonary infections (Parsek and Singh 2003). Many of us will have "the old man's friend" (bacterial pneumonia) living right with us as we approach the ends of our lives.

3.6
The Human Digestive System

If we include the oral ecosystem in the digestive system, this reinforces our contention that this organ system constitutes a bacteria-laden food conduit over which the human body exercises a measure of control in various regions by various means. The intake into the system is not sterile, and was even less sterile in critical times when our race was in the early stages of survival and dominance on earth, so the success of this system must be assumed and it is useful to search for the reasons for this success. The concept of barrier populations that screen tissues from attack by pathogens is especially germane in the digestive system because, until the civil engineers among us separated our sewage from our drinking water, the end product of one person's tract was included in the intake of everyone's system!

The oral ecosystem has two major inputs that affect the microbial flora of this most nutrient rich of all organ systems, and these comprise the serous crevicular fluid that bathes each tooth at a surprisingly high rate and the voluminous product of the salivary glands that irrigates the whole mouth. The hundreds of bacterial species that respond to the nutrient opportunity of the mouth must deal with antibacterial factors (e.g., antibodies) in these fluids as they try to muscle their way into preexisting bacterial populations on very different available surfaces. The tongue has a huge surface area that often develops an often visible microbial biofilm among its papillae and clefts, but it has attracted little study because it impinges on human health only in those fastidious enough to worry about bad breath. Similarly, the populations on the buccal mucosa are enormous but little studied, while those of the teeth and the gingival crevice have been studied by thousands of labs because they cause the caries and the gum diseases that have (until recently) made most of us edentulous well before our lives' ends. Conservative estimates place > 400 bacterial species in the biofilms that extend from the crowns of our teeth right down to the depths of our gingival crevices, which may approach 11 mm below our gum lines in those of us who are "long in the tooth". The oral ecosystem appears to respond well to the adage, popular in biofilm circles, that regular mechanical removal of the most obvious and disgusting slimy masses favors good environmental health.

The use of culture methods to define the species-rich "aerobic" biofilm on supragingival surfaces of teeth has led to the nomination of *Streptococcus mutans* as the major villain in the development of caries that tunnel into the hydroxyappatite of the tooth (Marsh and Bradshaw 1995). We have examined this attack by growing biofilms of *S. mutans* on hydroxyappatite in situ, and examination by NMR microscopy has shown very little of the lactic acid that is supposed to mediate the actual chemical attack on this crystalline substrate. We can use vertical scanning inferometry (VSI) to locate nascent caries, and we are analyzing this population by D-HPLC, so that we will soon be able to use FISH probes to identify all species involved in all stages of caries development by natural mixed-species biofilm populations. We predict that *S. mutans* will emerge from these direct analyses as a member of a consortium of bacteria that live in biofilms on teeth surfaces and cooperate in the development of a region of high proton and high organic acid concentration that initiates a focused attack on enamel. If this is true, caries formation will join microbially influenced corrosion (MIC) of metals as examples of cooperative metabolic processes carried out by bacterial consortia within biofilms (Fig. 3) that facilitate their focused attack by providing stable juxtaposition of the partners and local concentration of their products (Costerton and Stewart 2001). We may want to delay the use of *S. mutans*-based strategies for caries prevention until we determine what (if any) bacterial partners are involved in this relentless assault on our dentition.

The basic rules of microbial ecology are even more applicable to subgingival biofilms, because conventional studies of gingivitis and periodontitis have produced a list of species that are usually present in disease, without clearly identifying any single putative pathogen (Lamont and Jenkinson 1998; Davey and Costerton 2006). As we realize that inflammation plays a very large role in the etiology of biofilm diseases, we can conceive of periodontal disease as the response of the gum tissue to the juxtaposition of microcolonies of various organisms that comprise the biofilm in this constrained area. A healthy gingival crevice would contain a biofilm whose members did not attack the gingival epithelium and whose component species elicited a controlled and appropriate immune response. If invasive species (Lamont et al. 1995) usurped memberships in the subgingival biofilm population, or if organisms that elicit a damaging inflammatory response assumed positions (Moter and Gobel 2000) close to sensitive tissues (Fig. 47), then a disease process would be set in motion. Even the most sanguine of conventional culture-based microbiologists will cheerfully admit that less than half of the exotic organisms that populate this very specialized ecological niche have been grown in culture and identified. If gum diseases flair up in proportion to the species present in the gingival crevice, perhaps the very first area in which we should employ

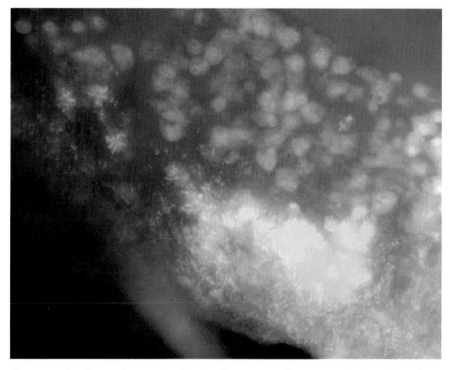

Fig. 47 Confocal image, in $x-y$ axis, showing the presence of matrix-enclosed biofilms of *Porphyromonas gingivalis* (*green*) and *Tanerella forsythensis* (*orange*) in the sulcus of a patient with controlled periodontitis. Polymorphonuclear neutrophils (PMNs), with blue nuclei, can be seen to be mobilized from tissues (*upper right*) that are mounting an inflammatory response to the presence of these biofilms. (Courtesy Annette Moter, Charite Hospital, Berlin)

the powerful new tools of DNA-based population analysis (D-HPLC) should be the gingival population. It is comforting to know that Mark Shirtliff will soon undertake just such an analysis, sponsored by Philips Oral Healthcare, and we look forward to welcoming and identifying hundreds of previously anonymous bacterial inhabitants of this cozy but "ticklish" ecosystem.

The human stomach produces prodigious amounts of hydrochloric acid, which has, until Barry Marshall's revelations in the 1980s, been credited with keeping this organ sterile while treading a fine line between bacterial control and tissue damage in the form of ulcers. In the first of many surprises concerning "sterile" organs and tissues, Barry discovered (Marshall and Warren 1984) that the stomachs of some individuals (and some family groups) harbor a burgeoning population of *Helicobacter pylori* that can invade the epithelium and cause ulcers by tissue invasion. While this beautifully adapted bacterium thrives in the mucus of the very acid stomach, the confluence where the acid-treated chime from the stomach meets the equally inhospitable bile-laden stream from the gall bladder produces a bacterial nightmare in the duodenum. The refreshing samples of stomach acid that rugby players often experience late in bibulous evenings give us some impression of the stringent conditions that bacteria in our food encounter when they enter this maelstrom and begin to drift down the stygian river of the duodenum. It is also worthy of note that the pyloris and the sphincter of odi both operate under the control of the autonomic nervous system, so that the disgusting mess that enters the duodenum is squirted into this proximal organ of the gut in a highly "processed" form.

The 10 m of gut that produces exquisite enzymes for digestion and has sufficient surface area for absorption would seem, at first glance, to expose the body to invasion by any bacteria capable of colonizing and invading the intestinal epithelia. But the intelligent design of the intestine produces a moving sheath of mucus (Fig. 48), at least 200 μm thick in all regions, that is propelled down the organ system by the peristalsis created by the smooth muscles of the gut wall. This moving mucus sheath confines most bacteria (Fig. 49), and most large food particles, to a de facto channel, so that only small molecular products of digestion can penetrate the mucus and contact the membranes of the microvilli. The efficacy of this strategy to limit the bacterial exposure of the intestinal epithelium is attested by the fact that it is difficult to produce animal model infections by simply feeding the cogent pathogens and the fact that ligation to limit peristalsis often facilitates infection (Caldwell et al. 1983). As in many cases, the ecological rule is clarified by its exceptions, and the human pathogens that succeed in infecting the gut all have special countermeasures to thwart the basic "mucus sheath" barrier. *Vibrio cholerae* has enjoyed enormous success as an intestinal pathogen by the simple expedient of producing a diffusible toxin that stimulates fluid release and washes the mucus off of the intestinal epithelium. *Shigella shiga* denudes the epithelium equally effectively by the release of toxins that cause arterial bleeding, and

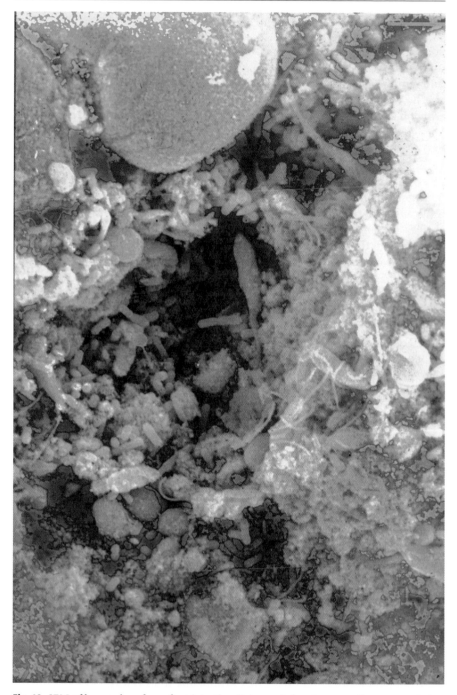

Fig. 48 SEM of lumenal surface of rat intestine (jejunum) showing very diverse population of bacteria and protozoa in mucus layer adjacent to microvillar surface of villi (*top*)

Fig. 49 SEM of same area of lumen of rat intestine showing morphology of some members of this fascinating microbial community, of which only a very few members have been identified or obtained in pure culture

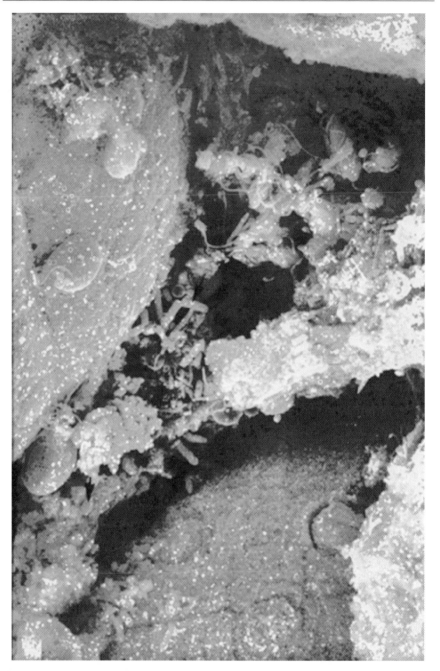

Fig. 50 SEM of same surface, in same animal, showing natural cleavage of mucus layer closer to tissue surface. Note the bacterial cells at the entrances of the crypts, the Giardia cells (one, at *8 o'clock*, detaching) on the villus on the *left*, and the three scars left by Giardia in the microvillary surface of the villus on the *lower right*

pathogens in the genus *Salmonella* are readily absorbed into the blood stream when some of their number can reach special receptor cells in the crypts between villi. Pathogenic Campylobacter have thick-sheathed flagella (Geis et al. 1989) that allow them to swim through the mucus and approach the intestinal epithelium, while pathogenic protozoa like Giardia (Adam 2001) can penetrate the mucus and adhere to the microvillar surface by means of specialized "suction cups" (Fig. 50).

The digestive tract provides a fascinating exposition of the strategies that plant and animal cells use to control bacterial colonization and invasion in natural ecosystems, and the efficacy of these strategies has allowed me to dine in bus stations from Rawalpindi to Peshawar. At the inlet of the tract, bacterial competition and the immune components of crevicular fluid and saliva make it very difficult for nomadic pathogens to integrate themselves into the biofilm populations that populate every specialized ecological niche. Food then passes rapidly down the esophagus to the hostile environments of the stomach and the upper duodenum, where only the most acid-resistant organisms (e.g., *H. pylori*) can survive to colonize and invade tissues and only very occasionally mount successful attacks up the bile duct and into the liver. This stressed bacterial population then embarks on its voyage down the mucus-sheathed intestine, propelled by peristalsis, and cells lacking specialized pathogenic mechanisms (i.e., most bacteria) would "see" the intestinal epithelium only very fleetingly, and only in the nether reaches of the system. In the event that any ambitious bacteria should have the initiative and temerity to approach and begin to colonize any part of the distal intestinal epithelium, the human gut reserves the right to invoke hypermobility and propel the offenders unceremoniously from the premises! So, while the human gut contains bacterial cells in all of its contiguous regions, it invokes a brilliant series of variations on the basic barrier theme to control colonization and invasion by all but the most specialized pathogens.

3.7
The Human Ecosystem: an Emerging Perception

While chemistry is subdivided on the basis of the types of molecules studied and the methods used, and biology is organized on the basis of the categories of living things being considered, microbiology is subdivided along anthrocentric lines. Even if microbiologists use very similar methods, and even if they study the same organism (e.g., *P. aeruginosa*), medical microbiologists and plant pathologists attend different meetings and read different journals, and new ideas permeate their disciplines at different rates. Because our science has concentrated on mitigating the effects of bacteria on human beings, medical microbiologists analyze specimens from diseased humans, while veterinary microbiologists tend to our sick animals, plant pathologists rescue our house plants and crops, and sanitary engineers protect our drinking water. Microbial ecology is a cross-cutting discipline that is not charged with the routine testing responsibilities that inevitably fall to the microbiologists who guard our health, and its practitioners have had the leisure to wander the byways and ponder the ways of bacteria. Microbial ecologists were among the first to turn to molecular biology for population analysis because natural ecosystems contained communities of great complexity and remarkable metabolic efficiency that yielded only a few sad stragglers and misfits on culture. Microbial ecologists were the first to realize the predominance of biofilms in natural ecosystems, and their examinations of plants and animals revealed the presence of commensal organisms that promote health as well as pathogens that cause disease. Ecologists think in terms of microbial consortia whose function is coordinated with that of eukaryotic hosts, and they would have predicted the commensal human skin population and the hardy creatures that inhabit extreme environments like the stomach and the bile duct.

At the recent conclave of microbial ecologists (ISME-11) in Vienna, in August 2006, the 2000 attendees were offered a session on microbial ecology and human health in the same meeting that featured the newest methods for in situ examination of intact microbial communities. People interested in human systems saw Mickey Wagner's and Holger Daim's brilliant use of FISH probes to identify dozens of species that form metabolic consortia in biofilms and their MAP method that feeds radioactive substrates to the whole community and assesses the metabolic activity of individual community members. They saw the synchrony of bacteria with eukaryotic hosts, ranging from sponges to dolphins, and discarded the archaic notion that the human body is virtually sterile until it is invaded by "dedicated" pathogens. They absorbed the ecological concept that a microbial community may function poorly if some of its members (e.g., *Clostridium difficile*) come to unnatural predominance or if some of its key members are killed by broad-spectrum antibiotics (e.g., bacterial vaginosis). This gradual refocusing of our attention on "ecological diseases", in which there is no single putative pathogen but there is an ecological shift away from healthy coexisting biofilms, has brought us to a point where we can begin to address otitis media (Hall-Stoodley et al. 2006) and sinusitis. Annette Moter stunned the audience with her direct visualization of the biofilm communities in the human gingival crevice (Fig. 47) and on infected heart valves, and we recoiled in horror at the festering mess that develops when our inattention to dental hygiene allows inflammation-causing communities to invade the firm pink gums of our youth. Microbial ecology has the potential to change medical microbiology beyond recognition because these ecologists approach the relationships between bacteria and the human body with a new mindset based on their studies of other animals and of plants. Bacteria obviously hold human beings in no particular awe or respect, and they are hard-wired for survival and dominance, so the microbial ecology of the human body is remarkably similar to that of any other ecosystem.

4 Replacement of Acute Planctonic by Chronic Biofilm Diseases

The acute bacterial diseases that bedeviled mankind until the middle of the last century (de Kruif 1926) constitute a biological anomaly. In these diseases, a bacterial species developed special properties that allowed it to bypass innate defense mechanisms and invade the human body, and then it reproduced very rapidly and used specialized toxins to kill the host before acquired immunity blocked its activities. In order to exploit the fleeting opportunity presented by naive human beings, specialized bacterial pathogens adopted the planktonic phenotype and produced millions of almost identical cells that charged through the tissues spewing enzymes and toxins. Unlike viruses, which mutate with dizzying promiscuity, the bacteria trotted out the same pathogenicity islands that produced the same toxins and mounted attacks on the human citadel with a predictability that could be countered with vaccines and antibiotics. "Here come the Goths (again); let's heat up the good old boiling oil!" Then "let's heat up the oil in case the Goths come again"! The human species was lucky, because the only time in the development of bacterial populations when diversity is sacrificed for reproductive expediency is during the exponential burst of growth that follows their discovery of an unprotected ecosystem. The two notable instances in which bacteria find these Shangri Las are in test tubes filled with fresh media and in naive animals that have not seen these particular bacteria in recent immunological memory.

These specialized pathogens could not sustain their frontal attacks on the human race because human survivors were immune, so most of them found reservoirs where they could withdraw from human contact and in which they adopted biologically correct less aggressive strategies. We have recently traced cholera, that faithful partner of the second horseman of the apocalypse (War), to innocuous biofilms in aquatic habitats where it lives in ecological harmony for decades, before it constructs its toxin machinery in response to temperature changes (Colwell and Huq 2001) and comes ashore to wreak havoc. Even the first horseman (Pestilence) hides from humankind by infecting fleas on rats, in a manner that doesn't kill the flea or the rat but produces a biofilm plug in the throat of these hapless insects that must be disgorged before the flea can feast on the blood of a human victim (Hurd 2003). So pathogens like *Vibrio cholerae* and *Yersinia pestis* are fine-tuned, both for survival in their reservoirs and for their entry into the human body, where the latter uses the flea as a vector and the former produces a toxin that stimulates fluid release and flushes the protective mucus from the victim's gut. Once they have gained access to the human body, specialized pathogens invade tissues as fast-moving planktonic cells, and time is "of the essence" because they will themselves be killed if the host was still alive to produce antibodies 7 to 10 d after the initial invasion.

As microbiologists came to understand the tactics employed by specialized pathogens, we were able to prevent epidemic diseases by counteracting their entry strategies, and the chlorination of water and control of parasitic insects improved our life expectancy. But the fatal element in the biologically flawed strategy of acute pathogens was their adoption of the planktonic mode of growth in the host. This mode of growth was duplicated in test tubes by Robert Koch (1884) and his systematic friends in Berlin, and it soon became obvious that planktonic bacterial cells are readily killed by simple sterilants and by antibiotics. Immunologists soon discovered epitopes on the surfaces of planktonic bacterial cells that could be used to produce vaccines that stimulated specific antibody production before the individual encountered the pathogen, and the era of the acute epidemic infections began to draw to its conclusion. Alexander Fleming noted that fungi have the ability to synthesize antibiotics that kill planktonic bacteria with gratifying speed and precision, and we gained the ability to treat acute infections that we could not prevent by vaccination. The harbingers of the vast array of specific antibiotics that now occupy whole walls in our pharmacies saved millions of lives during and after the Second World War, and new antibiotics soon sounded the death knell for acute epidemic bacterial infections. Medical microbiology has won a great victory over planktonic cells of specialized pathogens, and bacterial epidemic diseases are now mercifully rare in the developed world. Most of us will, however, actually die from acute infection by planktonic bacteria when age or physiological compromise sufficiently weakens our defenses, or when we confront antibiotic-resistant strains that we have produced in our quest for safe sterility, so the victory is not complete.

As the dramatic burden of acute epidemic bacterial diseases was gradually removed, and we no longer had to fear the deaths of our children from diphtheria or typhoid fever, a baseline of "low-grade" bacterial infections began to be recognized. Children were admitted to hospitals with middle-ear infections, women were affected with urinary-tract infections, and older men suffered the discomfort of prostatitis, but we could not isolate and identify a single specialized bacterial pathogen. Individuals were in acute discomfort from what appeared to be an acute bacterial infection, but cultures were only sporadically positive and yielded only a variety of "environmental" organisms, and frustrated clinicians even toyed with notions of a "viral etiology" in otitis media. The antibiotics used to treat virtually all of these infections often alleviated overt symptoms, but similar exacerbations recurred at varying intervals and the disease often entered a chronic cycle in which quiet periods alternated with acute episodes. Vaccines were generally ineffective in the prevention of these "indolent" infections, and the euphoria of the conquest of bacterial disease began to fade away. While the shift from acute to chronic bacterial diseases (Donlan 2001; Costerton et al. 2003) was happening, in the second half of the past century, medicine and dentistry began an unparalleled initiative in the placement of a bewildering variety of plastic and metal devices to improve our quality of life. When devices like artificial hip joints were totally implanted in the body, there were no bacterial sequellae unless there was a failure of asepsis, and gradual improvements in operative technique reduced the rate of device-related infections to < 0.02% in well-managed facilities. But infection rates were much higher when devices traversed the skin, or other epithelial tissues, and antibiotic therapy was relatively ineffective in resolving bacterial infections of either implanted or transcutaneous devices (Fux et al. 2003). While we were struggling with these device-related and other chronic bacterial infections, they burgeoned to the point where they now constitute fully 65% of the infections treated by physicians in the developed world (Costerton et al. 1999).

When we demonstrated that bacteria in natural and engineered ecosystems grew predominantly in biofilms (Costerton et al. 1978, 1987), perceptive members of the infectious disease community offered to help in examinations of device-related and other chronic infections. Allan Ronald (University of Manitoba) and Tom Marrie (University of Alberta) arranged for us to recover devices that had become foci of bacterial infections and mobilized clinical data to produce publications with pivotal impact in medical microbiology and infectious disease. We described a very well-developed *S. aureus* biofilm (Fig. 51) in which bacterial cells had survived 6 weeks of very high dose antibiotic therapy (Marrie et al. 1982), and we found that the bacteria associated with infected urinary catheters (Fig. 52), vascular catheters (Kowalewska-Grochowska et al. 1991), and colonized endotracheal tubes (Sottile et al. 1986) grew in extensive and exuberant slimy communities. We then proceeded to find biofilms on a huge variety of failed medical devices including vascular catheters (Marrie and Costerton 1984; Reed et al. 1986; Raad 1998), peritoneal catheters (Dasgupta et al. 1987), contact lenses (Feldman 1992), orthopedic devices (Gristina and Costerton 1984), and mechanical heart valves, and a consensus gradually developed that device-related infections were caused by bacteria growing in sessile communities. Direct examinations of material from chronic non-device-related infections like dental caries (Gibbons and van Houte 1975; Kolenbrander and London 1993), cystic fibrosis pneumonia (Fig. 53) (Lam et al. 1980), osteomyelitis (Fig. 54) (Mayberry-Carson et al. 1984; Lambe et al. 1991; Shirtliff et al. 2003), prostatitis (Fig. 55) (Nickel et al. 1994), endocarditis (Fig. 26) (Sullam et al. 1985), and otitis media with effusion (Fig. 56) (Post 2001; Dohar et al. 2005; Hall-Stoodley et al. 2006)

Fig. 51 SEM of biofilm formed by *S. aureus* on the tip of a pacemaker lead that had become colonized secondary to an acute bacteremia. The preparative method for SEM necessarily involves dehydration, which condenses the biofilm matrix, but the bacterial cells can be clearly resolved and the remnant of the matrix can be seen where the cells make negative impressions (*arrow*)

Fig. 52 SEM of a biofilm that formed on the luminal surface of a Foley urinary catheter, showing the presence of a mixed bacterial population (cocci and rods), and nascent crystals of urinary salts that often occur in these communities

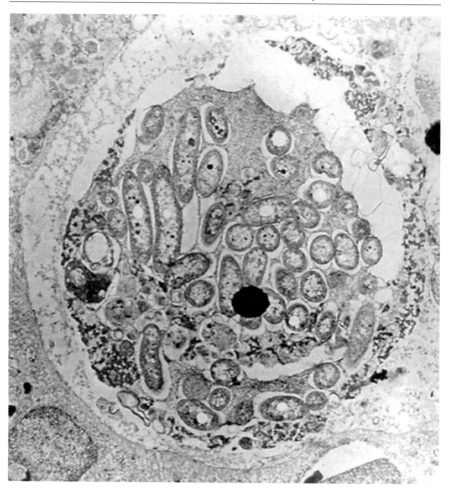

Fig. 53 TEM of microcolony of *P. aeruginosa* cells in alveolar space of a rat in which a model cystic fibrosis infection had been maintained for 23 d. The bacterial cells are seen to be enclosed in a matrix, which has been partially condensed by dehydration, and the microcolony is seen to be surrounded by an electron-dense "crust" that reacts very strongly with anti-rat-IgG antibodies

showed a similar preponderance of bacteria in biofilms. Now that we recognize a pattern of chronicity and recalcitrance in chronic bacterial infections, it is tempting to extrapolate and label a particular disease (e.g., chronic sinusitis) as a biofilm infection, but the best basis for this assignment is still the demonstration of sessile slime-enclosed aggregates by direct microscopy.

In the three decades during which we showed the presence of biofilms in almost the entire gamut of device-related and other chronic infections, the clinical earmarks of these infections were increasingly explained in terms of the inherent characteristics of bacterial biofilms (Hall-Stoodley et al. 2004).

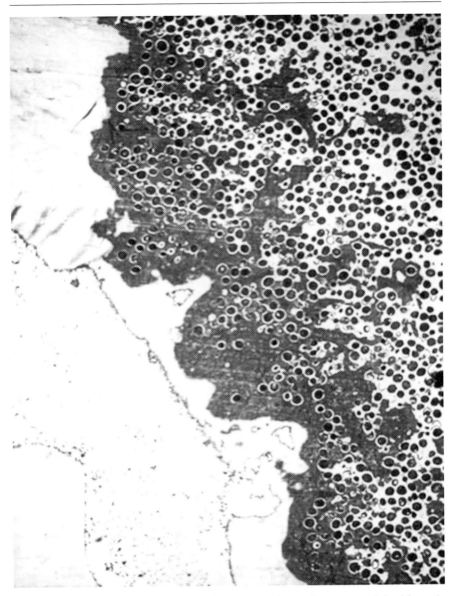

Fig. 54 TEM of prodigious biofilm that forms on surfaces of sequestra of dead bone in rabbit model of chronic osteomyelitis induced with *S. aureus*. Note the very large number of coccoid Gram-positive cells enclosed in matrix material that is well preserved near the bone and condensed elsewhere

We found that cells growing in biofilms are resistant to conventional antibiotics, at concentrations hundreds of times those that kill planktonic cells of the same strain (Nickel et al. 1985), and we attributed this inherent resistance

Fig. 55 SEM of bacterial cells in material removed from infected human prostate gland showing presence of very large numbers of two different sized rods clustered in patterns that suggest biofilm formation

to an altered phenotype found in sessile cells and in "persisters" (Lewis 2001). We found that bacterial cells in biofilms can grow and thrive in the presence of large concentrations of antibodies directed against epitopes on their surfaces (Fig. 53), and that biofilms can withstand the attack of activated phagocytes (Jensen et al. 1990), even in the presence of opsonizing antibodies (Fig. 24). We found very extensive microbial biofilms (Figs. 57 and 58) on invasive devices (e.g., Hickman catheters) that had never become foci of infection, and we described massive *S. epidermidis* biofilms on Tenckhoff catheters in subcutaneous tracts that showed no overt sign of inflammation. We noted that many toxin genes are turned off when bacteria grow in the biofilm phenotype,

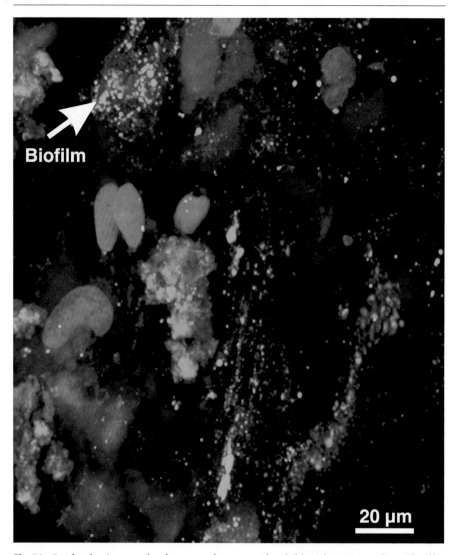

Fig. 56 Confocal micrograph of mucosa from ear of a child with otitis media with effusion (OM-E). Note the *orange* autofluorescence of the nuclei of the human cells and the presence of large numbers of bacterial cells that are *green* because they are alive, and the unfixed and fully hydrated preparation has been stained with the BacLite live/dead staining procedure. In the *upper left* quadrant (*arrow*) we see bacterial cells enclosed in matrix material, and we note that bacterial cells in infected tissues often display a wide range of sizes and shapes that differ from the very uniform dimensions seen in cultures. (Courtesy Paul Stoodley, Luanne Hall-Stoodley, Chris Post, and Garth Ehrlich)

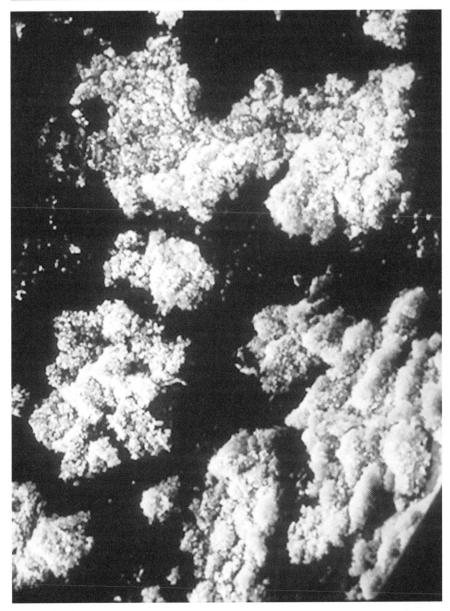

Fig. 57 SEM of luminal surface of a Hickman catheter that had been in place, in a cancer patient, for 3.5 months. Note that > 50% of this surface is covered by a barnaclelike mixed-species biofilm, within which coccoid bacteria appear to predominate

but that toxin production is reinitiated when individual cells are released from biofilms and assume the planktonic phenotype. In short, we gradually explained the salient characteristics of device-related and other chronic bac-

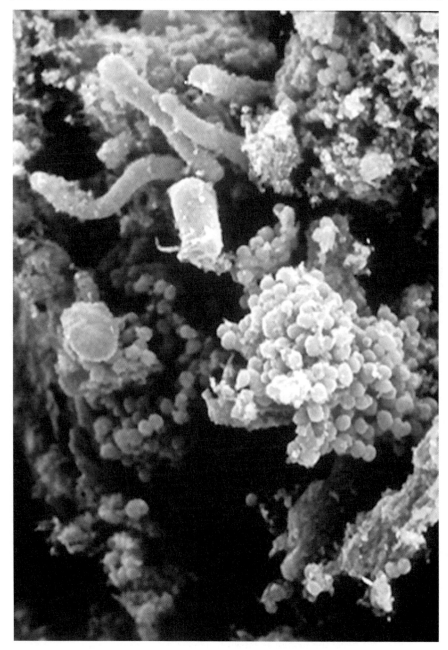

Fig. 58 SEM of a small part of a 1.8-mm mass of microbial biofilm that had partially occluded the cardiac tip of a Hickman catheter that had been in place, in a cancer patient, for 3.5 months. Coccoid bacteria and hyphal fungi are seen in this biofilm, and cultures from this device yielded *S. epidermidis* and *Candida albicans*. This patient did not experience any episodes of bacteremia during his 3.5 months of chemotherapy

terial infections in terms of the properties of bacterial cells growing in biofilm communities. Figure 59 contrasts acute and chronic infections and explains the properties of the latter in terms of the inherent characteristics of biofilms. Panel A shows planktonic cells that move aggressively through tissues but are susceptible to antibiotics antibodies and phagocytes. Panels B–D show

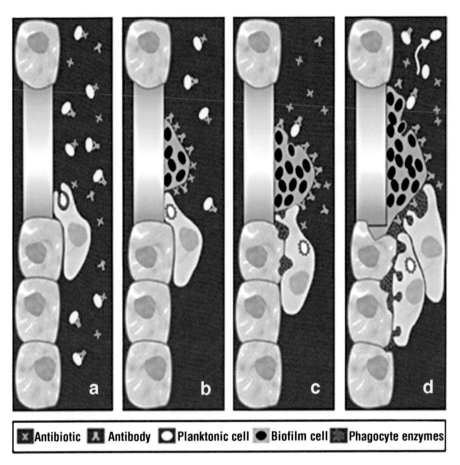

X Antibiotic **Y** Antibody ⬭ Planktonic cell ● Biofilm cell Phagocyte enzymes

Fig. 59 Conceptual drawing describing the essential differences between acute planktonic and chronic biofilm infections. *Left panel*: planktonic bacteria (*white*) are killed by antibiotics, antibodies, and phagocytic cells. In the next *panel* the bacteria have been converted to the biofilm phenotype (*black*), which provides protection from these antibacterial factors. In the next *panel* the biofilm is shown to release a planktonic cell, which is killed by an antibody, and a phagocyte is seen to confront the biofilm and to release degradative enzymes when its attack is frustrated. *Right panel*: a mature biofilm is seen to release sufficient planktonic cells to initiate an acute infection, and the frustrated attack of several phagocytes is seen to cause collateral damage to surrounding tissues. (Costerton et al. 1999; courtesy Science)

the gradual development of biofilms that are inherently resistant to these antibacterial agents and generally less aggressive in causing immediate tissue damage but potentially dangerous because they can stimulate inflammation and because they can serve as foci for acute exacerbations when they release planktonic cells. We did not set out to rationalize the differences between acute and chronic infections, but as we gradually came to understand bacterial biofilms, the properties of these sessile communities emerged as cogent explanations of the salient characteristics of the recalcitrant infections that they cause.

As we ponder the whole range of associations between bacteria and multicellular eukaryotic organisms, from simple commensal colonization to raging acute infections, a certain intellectual symmetry emerges if we adopt an ecological perspective. Both the eukaryotic organism and its bacterial partner obviously (and by definition) benefit from commensal relationships, and simple colonization by biofilm bacteria does not necessarily result in significant damage to the host. When we saw the extent of biofilm formation on the cardiac portion of the Hickman catheter shown in Figs. 57 and 58, we were horrified, but we noted that the patient had not experienced a single episode of bacteremia during the 3.5 months that this device had been in place (Tenney et al. 1986). Similar biofilm colonization (Fig. 60) of Techkhoff catheters, in the sensitive peritoneal region that responds so aggressively to planktonic bacterial invasion, produced no symptoms in large numbers of CAPD patients who used this system for years (Dasgupta et al. 1987). We must conclude that effective host defenses can "contain" the planktonic cells that detach from all biofilms, in most cases where devices are colonized by these bacterial populations, and that this colonization does not always result in damaging inflammatory reactions. Bacteria can then be seen to have established the ideal predator/prey relationship, in which the invading organism lives in a favorable ecological niche, for extended periods of time, because it avoids causing the cataclysmic demise of the host.

Growth in biofilms reduces the production of toxins, and fewer epitopes may be exposed at the surfaces of these communities, but tissue damage and inflammation are proportional to the size of the sessile community and extensive biofilms will usually inflict damage and elicit inflammation. Both of these processes may be as dependent on the species makeup of the biofilm, as well as on its extent, but very large devices (e.g., the Jarvik Heart) are easily colonized and the consequences are uniformly disastrous (Gristina et al. 1988). Between a simple suture and a huge hydraulic device that penetrates the chest wall there is a spectrum of surface areas occupied by biofilms, of tissue types impacted by the colonization, and of species compositions that may affect host reactions. Because of these multiple variables, manufacturers and users of medical devices are best advised to conduct early realistic animal experiments to determine the extent and species composition of colonizing biofilms, while designs can still be modified. Bacteria that have been declared

◀ **Fig. 60** SEM of the biofilm on the outer surface of a Tenckhoff catheter that had been removed (very carefully) from the abdomen of a CAPD patient in whom it had been in place for around 2 years. Note the complete occlusion of the surface of this device by a microbial biofilm, which is cracked because of dehydration during preparation for SEM. Large numbers of phagocytes are seen "patrolling" the surface of the biofilm, and they must have killed detached planktonic bacteria very efficiently because the patient did not experience peritonitis prior to catheter removal following a successful kidney transplant

to be nonpathogenic (e.g., *Pseudomonas fluorescens*) may be very damaging if they are introduced into the human lung in the form of biofilm fragments, and even *S. epidermidis* can be pathogenic in noncutaneous systems, so all bacteria are significant in colonization studies. In the dental area, biofilms in periodontitis and in colonized root canals affect local tissues, but they also have the potential to affect general health through the dissemination of planktonic bacteria and through the cumulative effect of inflammation involving large areas of affected tissues. Basically, the biofilm mode of growth enables bacteria to colonize and persist in mammalian tissues and organs, and the natural dynamics of this association means that many of us carry a microbial Trojan Horse from which destruction can emerge if our natural defenses fall below a certain threshold.

4.1
Etiology and Characteristics of Biofilm Infections

The mechanism by which bacterial biofilms damage human tissues and organs is the antithesis of the acute bacterial infections caused by planktonic cells of specialized pathogens. Most of this scurvy collection of microbiological misfits are simply organisms that are adapted to growth in the human environment (e.g., *P. aeruginosa*), or on human tissues (e.g., *S. epidermidis*), and are uniquely capable of protecting themselves from human host defenses. Because they are ubiquitous and prolific, they continuously challenge the human body and may gain entry at a compromised site as simple as a hangnail or as complex as a lung affected by cystic fibrosis. Most of these probing attacks are unsuccessful because the planktonic cells involved in these raids are susceptible to innate (e.g., defensins, surfactants) and acquired (e.g., bactericidal and opsonizing antibodies) defenses (Fig. 59), but it only takes one success to initiate the disease process. The establishment of a beachhead in human tissues is favored by the presence of inert surfaces because they are less well defended than the surfaces of living tissues, and the invaders can establish a defended biofilm on such surfaces in a matter of hours (Ward et al. 1992) if they are not challenged and killed. The presence of organic residues on inert surfaces is particularly invidious because it accelerates adhesion and biofilm formation, and metals, plastics, and devitalized bone are all favorable substrates for bacterial colonization and biofilm formation. Compromised tissues are also favorable for colonization, and elevated levels of salt in cystic fibrosis and glucose in diabetes favor colonization and biofilm formation.

The most common route of colonization is, however, an ecological shift in a commensal bacterial population that allows organisms to which the tissues are not adapted to proliferate and form biofilms on its surfaces. These ecological shifts in commensal populations may be gradual, as when the gingival crevice gradually becomes overgrown and anaerobic due to poor dental hygiene or the prostate becomes more heavily colonized because of an age-related deterioration in urodynamics. Similar ecological shifts may be episodic, as when vaginal and urinary populations change as a result of sexual activity, or they may be induced by the use of broad-spectrum antibiotics. Viral infections may cause abrupt changes in the colonization of tissues by commensal populations of bacteria, and very radical repopulations have been noted following viral infections in the human oropharynx (Harford et al. 1949). The result of all these ecological changes is that the commensal biofilm population, to which the tissue has become fully adapted, is completely or partially replaced by a community containing organisms with which the tissue is "less familiar". The bacterial species that replace commensal species in tissue-associated biofilms do not fit the classic "pathogen" rubric, and they would not satisfy Koch's postulates (Grimes 2006) even if they could be recovered in culture, but they become persistent members of the microbial biofilms aposed (Fig. 47) to the surface of the tissue.

To those of us who are trained in classic medical microbiology, the notion that a disease process is initiated by a nebulous population shift and that no single species can be identified as the "pathogen of record" is unfamiliar and exceedingly frustrating. But it is a modern fact of life and a strong indication that biofilm diseases are becoming predominant. We retain our ingrained habit of studying the "pathogen of record" so dental microbiologists study *Porphyromonas gingivalis*, and *Actionomyces actinomycetemcomitans*, and *Treponema denticola*, and *Fusobacterium nucleatum*, and *Streptococcus gordonii*, and other minor villains, but we still cannot nominate a single species (or even a constant cluster of species) that actually causes adult periodontitis. Research in otitis media is equally ambiguous, with nontypable *Hemophilus influenzae, Moraxella (Branhamella) catarrhalis*, and *Pseudomonas aeruginosa* vying for the nomination, and the stakes in prostatitis are simply wide open! If we pause to consider that we cannot identify a single pathogen, or a tight cluster of pathogenic species, in these three very common modern bacterial diseases, we are virtually forced to explore the paradigms offered by microbial ecology. We should probably analyze the commensal populations of the gingival crevice, the middle ear, and the prostate, and then we should examine the population shifts that have occurred in overt disease in these systems and try to determine which interlopers or responsible for the pathology. Culture methods are clearly unsuitable for this task, but a combination of molecular techniques for population analysis (Amann et al. 1995) and direct microscopy for species identification (Moter and Gobel 2000) should get the job done, and the ecological approach virtually demands this type of research.

The nidus of a biofilm infection may consist of a single-species community derived from a small number of planktonic cells that breached aseptic protocols during the installation of a medical device (Fig. 61), or it may be a complex biofilm containing dozens of commensal and extraneous species (Fig. 47). The latter extreme is seen in chronic wound infections like those treated in Randy Wolcott's very effective clinic (Lubbock, TX), in which contiguous sites along a surgical incision may display varying degrees of closure and truly revolting pus formation. The analysis of microbial DNA from pus from a single site, by the DGGE technique, regularly yields as many as 22 bands that indicate the presence of at least 22 different bacterial species (Fig. 62), while cultures yield only *S. epidermidis* and (occasionally) *S. aureus*. In general terms the species identity of the organisms that comprise a pathogenic biofilm is less important than the equivalent identification of

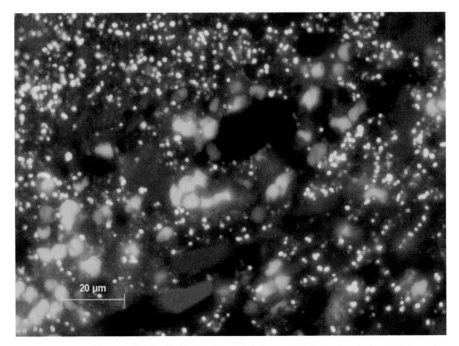

Fig. 61 Confocal micrograph of the surface of the aortic valve of a patient with native valve endocarditis that never yielded a positive blood culture. The preparation was reacted with specific FISH probes for viridans streptococci, and the infecting organisms can be seen, in typical streptococcal "chains" throughout the vegetations. (Courtesy Annette Moter, Charite Hospital, Berlin)

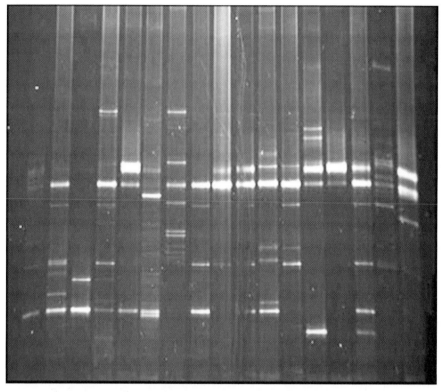

Fig. 62 DGGE gels produced using denatured bacterial DNA from chronically infected wounds in 17 different patients in Randy Woolcott's Southwest Wound Care center in Lubbock, TX. The multiple bands in each gel each indicate the presence of as many as 16 distinct species of bacteria, but specimens sent to a clinical lab have only yielded positive cultures for *S. epidermidis* and (occasionally) *S. aureus*

a planktonic pathogen, because directed antibiotic therapy is most unlikely to be effective. The microbial communities that form the niduses of millions of device-related and other chronic bacterial infections each year are consistently attacked by intact host defenses and by physician-directed therapy, and we can learn many salutary lessons by analyzing the outcomes.

The innate defenses of the human body are remarkably effective in removing potentially pathogenic biofilm niduses in certain organ systems. The squamous epithelium of the bladder sloughs when it becomes colonized by bacterial biofilms, so this otherwise vulnerable organ system is effectively "self cleaning", until the sustained presence of bacteria leads to the formation of intracellular biofilms in the "coffee pod" structures seen by Scott Hultgren's group (Anderson et al. 2003). Biofilm niduses in the peritoneum and the alveolar areas of the lung are walled off, by fibrosis and subsequent calcification (Fig. 46c), so that the bacteria are isolated and incapable of mounting an acute infection. Infected devices are often marsupialized, so that they are isolated from unaffected tissues, and they may even be externalized by this process. This natural process of removing biofilms is frequently assisted by surgery, and aggressive debridement of soft tissues is the standard of care in wound infections, while meticulous removal of all affected bone is demonstrably effective in the treatment of all forms of osteomyelitis. Extensive experience with infected hardware has led the orthopedic community (Costerton 2005) to conclude that colonized devices should be removed, sooner rather than later, to avoid continuing bone loss during repeated cycles of antibiotic therapy that relieve symptoms but fail to resolve the nidus. A simple methylene-blue-stained wet mount of a surgical suture removed from a patient's skin, as well as the most primitive of microscopes, will soon convince any clinician of the futility of trying to kill a healthy bacterial biofilm ensconced in a slimy matrix on an inert surface.

While observant physicians can learn valuable lessons from the dispassionate examination of clinical experience, classically trained microbiologists must reboot their mental hard drives. When I saw thick multispecies biofilms on the surfaces of Tenckhoff catheters (Fig. 60) freshly recovered from the peritoneum of CAPD patients, and equally luxuriant biofilms on the cardiac portions of Hickman catheters (Fig. 58) recovered from cancer patients, I assumed that these people must be critically ill. When I met patients who still had their catheters in place, and chatted with them while they strolled smiling down hospital corridors, I was inspired to erase my mindset to the effect that the presence of bacteria in the peritoneum or the heart is always disastrous. Biofilms are well tolerated in locations where an equivalent number of planktonic cells would be very dangerous. As we examine the proteins produced by bacteria growing in the biofilm phenotype (Fig. 23b, bottom), we note the absence of many toxins and enzymes that contribute to the aggressive attack of planktonic bacteria, and we speculate that this may explain the indolent nature of biofilm infections. The examination of more than 80 Tenckhoff catheters worn by CAPD patients for 6 months to 4 years showed the presence of remarkably thick and contiguous mixed-species biofilms (Fig. 60), but the patients reported no symptoms (Dasgupta et al. 1987) and most proceeded to successful kidney transplants. The major risk to these patients is acute peritonitis, caused by the release of planktonic bacteria from these biofilms, and the determining factor is the host's immune status and not the extent or species content of the biofilm.

If many biofilms are well tolerated, and if their sessile cells produce few if any toxins and enzymes, what is the actual mechanism by which they cause tissue damage? What tips the balance between biofilm colonization of the inert surface or the tissue and the infectious process that leads to tissue damage and pathology? The answer to these questions involves inflammation, and the first solid clues have emerged from work done by Curt Nickel in his elegant studies of the microbiology of the human prostate. Curt analyzed the bacterial population of the prostate glands of more than 300 volunteers, at Queen's University in Kingston, Ontario, and found that the extent of colonization and species makeup of these commensal biofilms were remarkably similar in all individuals. Curt then analyzed the type of inflammatory response mounted by these colonized individuals, in the age 55 to 60 age cohort, and discovered that all of the men with symptoms of bacterial prostatitis produced cytokines involved in the TH 2 response pattern. This extremely perceptive and pivotal observation removes much of the mystery from biofilm infections and suggests that tissue damage in these chronic infections derives largely from inappropriate inflammatory reactions to biofilms juxtaposed to tissues.

The new codicil to the biofilm hypothesis would, therefore, state that *the juxtaposition of bacterial biofilms to tissues that are not adapted to their presence may trigger deleterious inflammatory sequellae.*

In this very appealing hypothesis the role of the bacteria in the pathogenesis of chronic infections is simply to maintain their foothold (Fig. 59) next to susceptible tissues that are "programmed" to respond in a damaging pattern. Bacteria can cause chronic infections by forming monospecies biofilms on inert surfaces next to bone cells that will react by inflammation in all individuals, because bone cells are never adapted to the presence of any bacterial species. Similarly, bacteria can trigger damaging inflammatory responses in the lung or the peritoneum if they can persist in protected biofilms and disseminate sufficiently to avoid isolation and calcification. Bacteria can also join biofilms on tissues that are normally colonized by commensal populations and induce damaging inflammatory reactions if the tissue is not adapted to their presence. Because this inflammatory reaction depends on the prior "experience" of the tissue, and on the genetically determined nature of the inflammatory response, some individuals will suffer tissue damage while others will not. This hypothesis is particularly useful in understanding chronic infections of tissues (e.g., middle ear, prostate) that we now know to be colonized by extensive bacterial biofilms but in which symptoms appear in only some individuals. It may be useful to use this hypothesis as a framework to understand less studied chronic infections, like sinusitis, because these membrane-lined cavities are extensively colonized by biofilm-forming bacteria (Fig. 63), but damaging inflammation only occurs in some individuals.

The corollary of this inflammation hypothesis is that bacteria cause device-related and other chronic infections by the simple expedient of forming protected biofilms that allow them to persist in juxtapostion to tissues that are not adapted to their presence. Because the inflammatory response of the human is genetically determined, and highly variable, this perception introduces a host factor into the equation of susceptibility to chronic infection that is especially intriguing because it is potentially open to manipulation. When inflammation was first proposed as a major mechanism of tissue damage in chronic infections, Auerbach and his colleagues (Auerbach et al. 1985) used

Fig. 63 Confocal micrograph of material from the nasal passages of a patient with rhinos-inusitis, reacted with DAPI and a FISH probe specific for *Hemophilus influenzae*. Note the presence of a very large number of bacterial cells embedded in a cloud of clearly visible matrix material. (Courtesy Jeff Leid; see also Sanderson et al. 2006)

prednisone to reduce inflammation in the lungs of cystic fibrosis patients infected with biofilms formed by *P. aeruginosa*. But prednisone is a somewhat draconian immune suppressant, and pulmonary infections are potentially fatal, so the distinctly counterintuitive strategy of suppressing immune reactions in patients with bacterial infections should probably proceed very conservatively. Specific cytokines can be suppressed, to modify a TH 2 response to a TH 1 response, and diseases in which inflammation plays a large role in symptom production (e.g., otitis media and prostatitis) may be more suitable systems in which to test inflammatory modulation. As medical and dental treatments become more holistic and more tailored to the individual patient, we may begin to identify people who suffer from allergies and from inflammatory reactions to the presence of bacterial biofilms that do not affect their less reactive compatriots. My clinical friends tell me that they are haunted by miserable souls who progress from ear infections, to asthma, to periodontitis, to prostatitis, and I now believe that a more complete understanding of their immune systems may offer some resolution. The fault, Dear Brutus, is not in our stars, but in ourselves, that we are underlings!

If we focus on the direct evidence (Costerton et al. 1995, 1999) that device-related and other chronic bacterial infections are caused by bacteria growing in biofilms (Parsek and Singh 2003) and embrace the hypothesis that tissue damage in these infections results from inflammation, we can propose a new paradigm. The classic paradigm in medical microbiology considers specialized bacterial pathogens that invade organs and attack tissues with toxins and enzymes. We have traditionally defended ourselves with vaccines and antibiotics, and we have avoided contact with the pathogens concerned, but this has been notably unsuccessful in preventing or treating the current spate of chronic bacterial infections. If we grasp the new paradigm, we will use avoidance, immunization, and antibiotics to prevent and treat acute infections, but we will not use them to prevent or treat biofilm infections. Instead, we will manage tissue populations of commensal bacteria with the objective of avoiding ecological shifts that introduce organisms to which the tissues are not adapted. We will assiduously avoid the overuse of antibiotics that can produce dangerous changes in commensal bacterial populations, and we will explore the use of probiotics to reinforce and reestablish these beneficial biofilm communities when they are threatened. We will analyze the cytokine profiles of individuals who respond to the colonization of their tissues by mounting a damaging inflammatory response, and we will explore modern methods of manipulating this reaction to minimize this damage. We will attack biofilms on devices by prompt removal and active debridment, with aggressive antibiotic therapy to preclude recolonization, and we will limit the spread of biofilms in compromised organs (like the lung in cystic fibrosis) with sustained high-dose antibiotic therapy (Doring et al. 2000).

4.2
Biofilm-Based Strategies for the Prevention and Treatment of Chronic Biofilm Infections

The strategies that will protect us from device-related and other chronic bacterial infections must be interfaced with the strategies that have been so successful in protecting us from acute planktonic bacterial diseases. We must still avoid contact with specialized pathogens, we must still immunize our children, and we must manage our antibiotic armamentarium so that we always have effective agents to kill bacterial pathogens that overwhelm our defenses and cause acute infections. These measures are insufficient to protect us from the burgeoning spate of chronic biofilm infections that affect us in the developed world (Costerton et al. 1999), and we must design a new prophylactic strategy with these diseases in mind. Because commensal populations are critically important in preventing the invasion of extraneous bacteria, we must concentrate on the "care and feeding" of our microbial allies by good health practices and by avoiding broad-spectrum antibiotics wherever possible. We must also concentrate on the protection of temporarily compromised tissues, and of inert materials that make contact with tissues (e.g., contact lenses), to ensure that potentially harmful biofilms cannot develop and affect healthy tissues. The control of biofilm infections will not come from dramatic innovations like vaccination and the discovery of penicillin, which offered universal prevention and treatment for whole categories of infectious diseases. The control of biofilm diseases will come from improved management of compromised individuals and from the use of biofilm inhibitors (Fig. 35, bottom) and immune modulators in a manner that is tailored to the needs of the individual. Phase II of the conquest of bacterial diseases will be labor intensive.

Because most of the bacterial species involved in biofilm infections are virtually ubiquitous, it makes no sense for healthy individuals to avoid the biofilm-forming strains of *P. aeruginosa* that live in tap water and thrive in wilted plants. But the Pseudomonas invasion of the lungs of patients compromised by cystic fibrosis has been delayed by decades by the fastidious separation of colonized and uncolonized patients (Hoiby 2002) in the Copenhagen clinic directed by Niels Hoiby and Helle Johansen. As Niels and Helle report that one patient with the primary cystic fibrosis mutation remains uncolonized by *P. aeruginosa* at the age of 18 years, it is sobering to remember that cystic fibrosis patients were sent to special cystic fibrosis summer camps in the United States, as recently as 1988. But healthy individuals waste their time and energy if they try to avoid the biofilm-forming strains of Pseudomonas or of the *S. epidermidis* that populate our skin, or the equally gregarious *S. aureus* that live in all of our noses. One area in which biofilm avoidance does, however, pay dividends is in aerosol management, in that the inhalation of preformed biofilm fragments is to be avoided at all costs, since these protected aggregates can never be cleared from our lungs (Fig. 46). For this reason, air conditioning systems should be cleaned before they are activated in the spring and dental professionals must be very careful to wear effective masks and avoid direct aspiration of potentially dangerous aerosols.

With these few exceptions in mind, our time will be better spent in cultivating healthy lifestyles and avoiding intrusions that can give biofilm-forming bacteria a foothold in our bodies. If we floss our teeth, our gums will remain pink and anaerobic bacteria will remain under control in our gingival crevices, and we will avoid dental prostheses and root canals if we pay attention to our general dental health. Our pulmonary and digestive health will be favored if we remember that it is better to eat or drink biofilms than it is to breath them, and the commensal biofilms in our naughty regions will continue to protect us if we manage our social lives to minimize contact with microbial "strangers". The greatest threat to our commensal biofilms is the misuse of broad-spectrum antibiotics, and we should inform patients that this hazard weighs equally with the generation of resistant strains, among the reasons for restricting their use to acute bacterial infections. The realization that our immune systems control biofilm infections, by killing planktonic bacterial intruders and by mounting vigorous but appropriate inflammatory reactions to the presence of these microbial communities, should make us pay more attention to our general health. Simple intrusions into our tissues, like a sliver or a hangnail, should be seen as a reservoir of biofilm bacteria and must be monitored for signs of a generalized acute infection. The answers seem to lie in better motherhood and more apple pie!

4.2.1
Reduction of "Bacterial Loads" and Colonization Rates

Skillful surgery and fastidious asepsis have enabled us to extend good quality of life for decades by the replacement of worn-out joints with metal and plastic prostheses. In spite of our natural tendencies to emphasize our own areas of interest (e.g., infection), we should remember that the vast majority (> 99%) of deep implantations of such devices as prosthetic hips are successful and uncomplicated by bacterial colonization or infection. Good asepsis reduces the number of bacteria that have access to the operating field, and the surfaces of sterile plastic and metals are not inherently favorable to the survival of these essentially planktonic cells in the presence of surfactants and sufficient quantities of perioperative antibiotics. Tony Gristina accurately dubbed this phase of implantation the "race to the surface" (Gristina and Costerton 1984; Gristina et al. 1987), in which cells must survive long enough to accomplish the transition from the planktonic phenotype to the antibiotic-resistant biofilm phenotype, and to construct the matrix-enclosed biofilm within which they are protected from host defense factors. In most cases the doctors and patients win this race, but in some cases the bacteria win the race and initiate a process in which colonization may lead to overt infection in a few weeks or in several years. Many factors can tip the balance in favor of the bacteria in this critical race, and the experience of the US Navy is germane in that rates of marine fouling are much accelerated if the surface to be colonized bears the remnants of previous biofilms. We had a very disturbing experience in which a medical device was manufactured in a biofilm-rich environment and was sterilized but not cleaned prior to being implanted in patients, many of whom developed devastating infections. On reexamination the plastic surfaces of this device were covered with a layer of biofilm remnants about 35 μm in depth, and we concluded that this preformed biofilm matrix favored biofilm formation by planktonic bacteria and subsequent infection. Because most infections of certain types of medical devices involve certain bacterial species (e.g., *S. epidermidis* on mechanical heart valves), we can model this process of successful colonization and biofilm formation and predict factors that accelerate or delay this grim progression.

While implanted devices are only exposed to bacteria during the implantation process and during hematogenous episodes, transcutaneous devices are continuously exposed to bacteria that colonize their surfaces and produce biofilms of prodigious dimensions. Our direct studies of Tenckhoff peritoneal catheters (Fig. 60) and of Hickman vascular catheters (Figs. 57 and 58) have shown that more than half of both the luminal and external surfaces of these devices are covered by extravagant multispecies biofilms composed of both bacteria and fungi. Biofilm engineering has taught us that biofilms shed planktonic cells in predictable numbers (www.springer.com/978-3-540-68021-5: Movie 8), and clinicians in the field of peritoneal dialysis have concluded that some patients can handle those cells shed from the surface of these 10- to 12-inch (25- to 30-cm) perforated hoses, while others cannot and develop peritonitis. Hickman catheters of similar size were found to be equally heavily colonized by microbial biofilms when they were recovered after 2 to 3.5 months in place, but only 5 instances of bacteremia were recorded in 3 of 81 instrumented patients (Tenney et al. 1986). Larger devices, and especially devices with access to bacterial contamination like the Jarvik heart, are more prone to produce overt infection (Gristina et al. 1988), and reductions in size (as in the case of the implantable defibrillator) have been shown to reduce infections to acceptable levels. We cannot calculate the "acreage" of biofilm-covered surface that will shed enough planktonic bacterial cells to overwhelm the host defenses in any organ, but once this point has been reached, the question becomes academic because an acute disseminated infection will have been initiated.

Biofilm control is also an important issue in the management of medical devices that are implanted for long periods of time. The exit sites surrounding transcutaneous devices are accessible to environmental organisms, and the biofilms that they form on the device and on the tissue are subjected to a truly sickening reciprocating movement as the patient breathes and moves. New dressings have been developed in attempts to limit these biofilm infections – vancomycin is frequently (over)used for their control – and John Olerud's cohort of our CBE wound center team now seeks to develop cuffs of material that integrates with the dermis and blocks bacterial access to deeper tissues. Indwelling urinary catheters have proven so difficult to protect from colonization (Fig. 52) and subsequent infection that competent patients are usually encouraged to practice intermittent self-catheterization to avoid chronic biofilm cystitis and its attendant complications. Catheter lock solutions have been useful in the control of luminal biofilm colonization of transcutaneous devices, and the most effective (and draconian) must surely be the "Y set" that allows CAPD patients to sterilize the distal portions of their Tenckhoff catheters with bleach! When the human circulatory system is interfaced with water, across a dialysis membrane in hemodialysis, the water must be free of both bacteria and their pyrogenic fragments (e.g., LPS). Water systems that test within the specifications of 200 bacterial cells/liter are frequently colonized by extensive upstream biofilms that contribute disturbing levels of pyrogens as well as episodic bursts of planktonic cells. The adoption of the same water preparation systems used by computer chip manufacturers, coupled with biofilm control based on Intelligent Optical System's biofilm probe (Fig. 33), would certainly slow the relentless downhill slide of the hemodialysis patient. The biofilm community is interdisciplinary and highly interactive, and biofilm-trained engineers can solve many medical problems.

4.2.2
Immune Monitoring and Immune Treatment of Biofilm Infections

The human immune system monitors the whole body for bacterial interlopers, with remarkable efficiency, and we know that the biofilm phenotype of each bacterial species expresses surface components (Fig. 23) that differ from those of planktonic bacteria. Laura Selan has solved an important medical problem by identifying a glycoprotein (SSPA) epitope that is expressed on the surfaces of biofilms cells of *Staphylococcus aureus* and *S. epidermidis*, and by developing an ELISA test for anti-SSPA antibodies (Selan et al. 2002). The problem involved the synthetic vascular grafts used to replace segments of arteries that would simply fail and rupture, without any warning symptoms, if the sutures that connected the artery to the graft became colonized by *staphlycoccal* biofilms. If biofilms of either staphlycoccal species formed on the sutures, as happens in ca. 4% of such implants, the ELISA titer rises above the baseline established for each patient at the time of surgery (Fig. 64) and the graft can be replaced before it fails and the patient dies. We have engaged a brilliant orthopaedic surgeon/scientist, Charalampos Zalavras, in the use of this ELISA test to detect device-related infections in complex trauma repairs and in the use of antibody-based imaging to determine which elements of fixation systems are infected. We believe that this will be the first of many instances in which new microbiological techniques will provide data that will be used, directly and immediately, to guide decisions on the clinical management of patients (Costerton 2005).

While antibodies against biofilm bacteria may be useful in monitoring ecological shifts in commensal bacterial populations and in detecting the onset of device-related infections, they are very damaging if they react with the biofilm but fail to clear it. The "crust" of IgG antibodies surrounding the microcolony of *P. aeruginosa* cells seen in Fig. 53 are present in the form of immune complexes, and their formation elicits significant tissue damage (Cochrane et al. 1988). Inflammation and frustrated phagocytosis (Figs. 47, 59c, and 60) are major factors in chronic infections, and the pathogenic role of the biofilm may be simply to persist and cause havoc by thumbing its nose [sic] at the humoral and cellular immune systems. In organs that cannot be sterile, like the middle ear in children and the prostate in old men, the immune system must generate enough reaction to prevent acute tissue invasion,

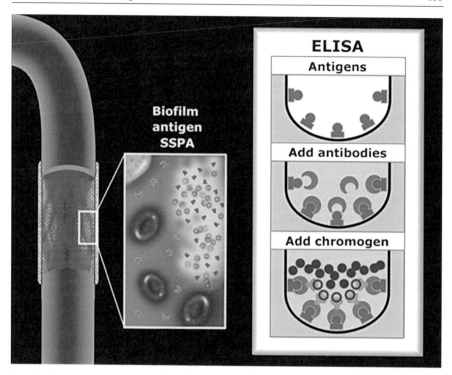

Fig. 64 Conceptual drawing of biofilm infections that sometimes occur when vascular grafts are used to repair arteries. If either *S. aureus* or *S. epidermidis* forms biofilms on the Dacron or Gortex grafts, their cells express the biofilm-specific epitope SSPA (*red*) and the patient produces anti-SSPA antibodies (*blue*). We measure IgM antibodies against SSPA in an ELISA test (*right panel*) and, if the titer rises in the 10 d following implantation, the graft is examined by direct imaging, and replacement is considered if the infection is confirmed by physical determinations

but it will damage the organ if it overreacts and produces cytokines that trigger inappropriate inflammation. This is the crux of biofilm infections, and the long-range solution may lie in the manipulation of the inflammatory process by the use of cytokines and cytokine antagonists. Jeff Leid of the University of Northern Arizona has examined material from inflamed sinuses (Fig. 63) and has found that these cavities are virtually filled with exuberant biofilms, in which a rich morphological variety of sessile cells thrive in a confluent matrix. Antibiotic therapy seems futile because biofilms are resistant and the sinuses will readily reinfect, surgery may rip up the face without offering benefits other than improved drainage, and inflammatory modulation may offer a reasonable and easily validated alternative.

Against this background of damage caused by immune overreaction and this talk of immune suppression, it may seem counterintuitive to wonder if a highly reinforced immune attack could really resolve a biofilm infection.

Sam Silverstein of Columbia University is the doyen of cellular immunology, and he is frankly dismayed by our conclusion that bactericidal antibodies and opsonin-enhanced phagocytosis are both ineffective in biofilm control. Sam has shown that mature biofilms can be resolved by activated phagocytes if the encounter takes place in the realistic milieu of a fibrin clot and if the attack of the phagocytes is reinforced by opsonizing antibodies and compounds that enhance the oxidative burst. We will examine the efficacy of attracting phagocytes into the areas immediately surrounding implanted devices and of tipping the balance in favor of these protective cells by providing opsonizing antibodies and compounds that enhance their oxidative burst. We will keep a weather eye on inflammation and its attendant tissue damage, but we will determine whether the human immune system can resolve some biofilms in some regions of the body if we can effectively tip the balance against the bacteria. Clearly, if anybody can do it, Sam can.

4.2.3
Direct Manipulation of Biofilm Formation by Signal Inhibition

Tony Gristina's "race to the surface" concept projects a useful image in which planktonic cells are susceptible to host defenses and to antibiotics (Fig. 59a), while sessile cells are protected from these exigencies as soon as they assume the biofilm phenotype (Fig. 59b). Factors such as bacterial number, surface cleanliness, and the concentrations of antibacterial agents will all affect the outcome of this race, but logic dictates that we explore all possibilities of keeping the bacteria in the planktonic mode of growth and of delaying biofilm formation. The gradual perception that all bacteria respond to a battery of "quorum sensing" signals (Singh et al. 2000; Fuqua and Greenberg 2002), and the recent realization (Davies et al. 1998) that biofilm formation is controlled by these signals, have raised the very real and practical possibility that device-related and other chronic bacterial infections can be prevented and controlled. Our primitive knowledge of the chemistry of the first few signaling systems to be defined has enabled us to construct specific competitive inhibitors (Fig. 35, bottom) of biofilm formation, and our surveys of natural ecosystems for compounds that protect plants from biofilm colonization have yielded dozens of potent biofilm inhibitors. The continuous presence of these biofilm inhibitors in aquatic ecosystems (McLean et al. 1997) argues against any toxicity for humans, and the thousands of years of efficacy of natural biofilm inhibitors argues against the emergence of bacterial resistance to this inhibition of biofilm formation. Promising biofilm inhibitors (e.g., inhibitors of biofilm formation by Gram-positive cocci) serve as bellwethers, because their efficacy in preventing biofilm infections are published (Balaban et al. 2003), but dozens of even more effective biofilm blockers are being secretly groomed by several well-supported commercial enterprises. We anticipate that biofilm inhibitors will be tested for their ability to prevent device-related infections, in large-scale clinical trials, within the next 3 years.

The scientific basis of infection control by biofilm inhibition has been meticulously laid by several insightful researchers with special talents for the design of unequivocal experiments. Mike Givskov has used signal reporter bacteria (*E. coli*) to show (Wu et al. 2000) that cells of *P. aeruginosa* growing in the lungs of mice both send and receive acyl homoserine lactone (AHL) signals, and that the administration of a specific AHL signal blocker (a brominated furanone) limits biofilm formation and allows the host-mediated clearance of these pathogens (Wu et al. 2004). Many refinements of these experiments will be done, and the effects of signal manipulation on natural bacterial populations will be pondered and assessed, but the principle that bacteria living in tissues send and receive signals and can be controlled by exogenous signal inhibitors has been unequivocally established. We must strain to grasp all of the ramifications of this new biofilm-based concept of chronic infections and the full significance of Givskov's incisive observations. Bacteria growing in biofilms in such chronic infections as otitis media and cystic fibrosis are sentient creatures, and we can "talk" to them by means of several classes of chemical signals. Some signals are so specific that we can chat with the *Pseudomonas* species without disturbing the other Gram-negative denizens of the human flora, while others (e.g., autoinducer II) are virtually universal (Schauder et al. 2001). Givskov's observations show that we can use signal inhibitors to influence bacterial behavior in such major matters as biofilm formation (Hentzer et al. 2002), but a much more subtle approach would be to use signals and inhibitors to control the growth rates of pathogens, or their production of inflammatory triggers, or their detachment from biofilms. As in all conversations, we must be very careful that communications with one "listener" do not cause untoward reactions in "bystanders" for whom the message is not intended, but the pivotal fact is that organisms that we previously conceived of as being deaf and mute can be contacted and manipulated by chemical communication.

In equally incisive experiments, Naomi Balaban and her colleagues have shown that biofilm formation by *S. aureus* in device-centered animal experiments can be prevented by the RIP inhibitor of the RAP signal that controls this process in this organism. When cells of *S. aureus* are injected into Dacron sleeves that have been implanted subcutaneously in rats, these organisms normally form biofilms and cause a chronic infection that cannot be resolved by antibiotic therapy. Local or systemic administration of RIP to these animals prevents biofilm formation on the Dacron material and allows the resolution of these model infections by host defenses aided by the use of systemic antibiotics (mupurocin), to the extent that no living pathogens can be recovered (Balaban et al. 2003). Because of their pivotal nature, these experiments have been repeated in several labs, and we must now consider the principles that have been unequivocally established by this work. Questions of toxicity and of effects on native bacterial populations will remain, and be resolved, but the rules of scientific discourse demand that we recognize or refute Naomi's claim that she can use a biofilm inhibitor to prevent biofilm formation by an aggressive pathogen in a device-centered animal model. Because of the virtual emergency caused by device-related infections in the developed world, where these infections now constitute the majority of bacterial diseases, granting agencies have issued specific RFAs for biofilm research (e.g., NIH PA-06-537) and regulatory agencies may decide to "fast track" biofilm inhibitors for the prevention and treatment of device-related and other chronic bacterial infections.

4.2.4
A Coordinated Approach to Biofilm Control

As we expand the tentative and timorous contacts between clinicians and microbial ecologists, through the mediation of the biofilm community, we may try to imagine ourselves in the situation of a bacterial cell that finds itself in the vicinity of a medical device during implantation. Microbial ecologists have the annoying but generally useful habit of trying to "think like" bacteria, and they perform many teleological contortions to explain why the vast majority of implanted devices do not accrete biofilms and become foci of infection. As in most real-world strategies, we simply intend to increase and broaden the bacterial "losing streak" by combining all possible factors that favor the host and hobble the pathogen in Tony Gristina's now axiomatic "race to the surface". As in all races, time is clearly of the essence, because planktonic bacterial cells can assume the biofilm phenotype and construct highly protected biofilms in less than 1 h (Fig. 22) after they find themselves at a surface in a permissive milieu. The condition of the surface is important, in that organic accretions will accelerate bacterial adhesion and consequent biofilm formation, and antibacterial agents (antibiotics, surfactants, antibacterial peptides, etc.) will kill planktonic interlopers if they are present in the fluids near the device in the prebiofilm time frame. It is highly unlikely that antibiotics will be effective in the fluids near the device, in the same order that they are effective in routine laboratory antibiograms, but we now have the necessary methods to assess their efficacy in animal tissues using the live/dead probe and the confocal microscope (Cook et al. 2000). Signal-based biofilm inhibitors, by definition, extend the period during which bacteria remain in the vulnerable planktonic state (Fig. 35, bottom), and their effects will be exponentially additive to the effects of agents that kill free-floating bacterial cells.

The major practical problem in the application of antibiofilm technologies in the protection of medical devices from infection has been an inexorable law of physics that dictates that the release of solutes from plastic coatings occurs rapidly at first and diminishes with time. Antibiotic-loaded plastics have been disappointing in infection control, but this problem may largely have been solved by a brilliant discovery in Buddy Ratner's lab at UWEB (www.uweb.uwashington.edu). Buddy's team has developed an ultrasonic-sensitive "skin" for plastics that retains solutes except when it is deranged by ultrasonic energy, at which time the solutes are released at high rates to produce high local concentrations (Fig. 65). Patrick Norris (CBE) has loaded hydromers with tobramycin and has shown (Norris et al. 2005) that the disruption of the UWEB "skin" by ultrasonic energy releases the antibiotic in a pattern that both precludes colonization by incoming planktonic cells and kills the cells in preformed biofilms. This "on-demand" technology will allow us to flood the immediate environment of newly implanted devices with any solute (e.g., antibiotic, biofilm inhibitor, antibacterial peptide) during and after implantation and at any interval following placement that is dictated by clinical considerations.

Biofilm engineers have discovered many ingenious ways of making biofilm bacteria "uncomfortable" and more susceptible to conventional antibacterial agents. Our group has used D.C. electric fields at low current densities (±2 mA) (Blenkinsopp et al. 1992; Costerton et al. 1994a; Wellman et al. 1996;

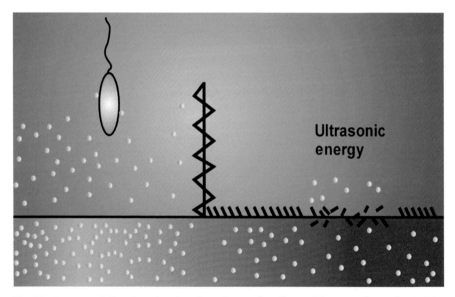

Fig. 65 Conceptual drawing showing the release of antibacterial agents from plastic biomaterials used in medical devices. If the agent is simply impregnated into the plastic, it will be released in concentrations sufficient to kill "incoming" bacteria (*left panel*) for a limited length of time, but it will eventually become depleted below effective levels. The UWEB center at the University of Washington (Seattle) has developed an oriented hydrocarbon "skin" that retains antibacterial agents until it is disturbed by ultrasonic energy (*right panel*), at which time the agent is released in a controlled manner. This technology can be used to protect medical devices from colonization by ultrasonic release perioperatively or when symptoms of infection are detected

Jass and Lappin-Scott 1996), and Bill Pitt (at BYU) has used ultrasonic energy (Nelson et al. 2002) at specific wavelengths to decrease the inherent resistance of biofilm bacteria to levels close to those of planktonic cells of the same strain. Surgeons are attracted to the notion that a device could be sterilized in situ, after implantation, by a physical treatment that enhanced the efficacy of intraoperative antibiotics, and endodontists would value an effective method of killing the bacteria that inhabit root canals before they are filled with guta percha. Surgery does not always involve the precise dissection of healthy tissues by pristine instruments, and many of my surgical friends actually find themselves cutting away at inflamed muscle in the vicinity of infected devices or at rotten nerves in teeth. In these cases, the antibiotics and sterilants that are prayerfully administered could be reinforced by externally induced electric fields and/or ultrasonic energy, and the chances of success could be improved. The fact that some medical devices are made of conductive metals and that most devices are structured in ways that enhance the effects of ultrasonic energy may favor adding physical insult to orchestrated chemical injury in our attempts to hobble the bacteria in the race to the surface.

In engineering terms, we will provide an accurate and impartial "test bed" in which antibiofilm strategies can be auditioned, alone or in an infinite variety of combinations, and we will use direct observation methods to see how many cells have survived and which species they represent. These direct methods will be used in a systems approach to the prevention and control of device-related infections, so that manufacturers can select agents that will be effective, but they can also be used to answer very simple and practical questions about these devices. For example, John Olerud (dermatologist at the University of Washington and father of Boston's stellar first baseman) has invented a cuff for vascular catheters that bonds with the dermis, and we are currently studying whether trapping bacteria below this "seal" is beneficial or deleterious. We will study wound healing and other device-related phenomena by building a "map" of the whole affected area, using FISH probes to locate bacteria in thick ($\pm 50\,\mu$m) frozen sections, and we will attempt to understand the role(s) of bacteria in enhancing or interfering with these processes. One question that recurs is whether an inherently commensal organism (e.g., *S. epidermidis*) can be beneficial in wound healing by building tertiary structures (Fig. 18) and by excluding overt pathogens by competition. This test bed will be established in our new NIH-supported CBE-UWEB wound center, and we have already set up a similar test bed in the Center for Biofilms (School of Dentistry, University of Southern California) for the assessment of root-canal sterilants. In this USC test bed we will remove teeth that cannot be saved by root-canal treatment, form and shape a root canal ex vivo, and assess the ability of dozens of sterilants (including the bleach solutions currently used) to kill the bacteria in the "smear layer" and the dentinal tubules. It may be imprudent to anticipate the results, but the fact that we have a test bed with accurate methods and solid metrics will surely allow us to select antibiofilm strategies that will reduce the presently unacceptable rates of device-related infections.

4.3
New Diseases, New Concepts, New Tools

Because the specialized pathogens that cause acute epidemic bacterial diseases depend on a frontal attack using toxins and a wide variety of invasive factors, they require an immunologically naïve host and they adopt the planktonic mode of growth in tissues. These acute diseases are easily and reliably detected by our standard 160-year-old culture techniques because planktonic cells are readily collected by swabs or in body fluids and they grow and produce a proportional number of colonies when spread on the surfaces of agar plates. These treasured techniques of classic microbiology can be practiced in even the most primitive surroundings, where acute diseases still predominate, and they have formed the perceptual basis for the early detection and virtual eradication of acute epidemic bacterial diseases in the developed world. But recovery-and-culture techniques have almost disappeared from the armamentaria of microbial ecologists working in natural ecosystems because we have concluded that we can culture only a very small proportion (< 1%) of the organisms we can see by direct microscopy (Greene 2002). As a consequence, recovery-and-culture techniques have largely been replaced by molecular techniques in microbial ecology, while they have persisted in diagnostic medical microbiology, despite their general failure to detect pathogens in device-related and other chronic infections.

This situation came to a head in an area of medical microbiology in which ecology and pathogenicity meet, in our studies of the colonization of the human female reproductive tract with toxigenic strains of *Staphylococcus aureus* that cause toxic shock (Veeh et al. 2003). Our commercial sponsors used state-of-the-art methods in the detection of *S. aureus* by swab-and-culture methods, in 3000 volunteers, and found this pathogenic species in 10.8% of the subjects. We then examined material from a subset of 300 of this same cohort, using 16 S rRNA-directed FISH probes for *S. aureus* (Fig. 41), and found very large numbers of these cells in all 300 subjects, and these results were confirmed by PCR techniques. We could find no relationship between culture positivity and the number of cells of *S. aureus* seen by the use of FISH probes, and culture positivity was seen to vary widely in swabs taken from the same individual at different times. We conclude that bacterial cells growing in biofilms simply fail to grow when recovered from natural and pathogenic ecosystems and when placed on the surfaces of agar plates (Maki et al. 1977) and that these classic microbiological methods only detect planktonic cells. Recent studies of "sterile loosenings" of Sulzer acetabular cups have shown the presence of *S. epidermidis* biofilms, by microscopy and by FISH probes, while culture methods have failed to detect bacteria in either the synovial fluid or on the surfaces of recovered devices from more than 1500 patients. This failure of recovery-and-culture methods to detect bacteria growing in biofilms has caused many perceptive clinicians to lose confidence in micro-biological data (Rayner et al. 1998), and it should prompt us to adopt new methods for the detection and study of device-related and other chronic bacterial infections.

Our realization that cells growing in biofilms fail to grow when they are placed on the surfaces of agar plates has been costly, in mistaken diagnoses and in missed therapeutic opportunities, but it comes at a very serendipitous time in terms of replacement techniques. The Mayo Clinic is currently processing all of its microbiology specimens by PCR, in parallel with conventional culture techniques, and companies are developing practical "kits" in which rRNA recovered from clinical specimens is reacted with 16 S rRNA probes to identify pathogenic species in body fluids. FISH probes are already used for the diagnosis of legionellosis (Hu et al. 2002), and molecular diagnostic techniques are coming into routine clinical use because they can detect biofilm bacteria and because they yield results rapidly enough to direct early therapeutic responses (Moter et al. 1998). We should accept the fact that culture methods will be used for several more decades, and we should realize what they actually mean. If bacteria are detected by recovery and culture methods, then they were actually present in the infected tissue in the planktonic form, and the antibiogram that accompanies positive cultures may be very useful in designing therapy to suppress the acute phase of the infection in question. But we cannot rely on this antibiotic-sensitivity information for the resolution of device-related and other chronic infections because these methods do not yield the biofilm-killing dose, and we must remember that negative results do not indicate the absence of pathogens growing in biofilms.

However, it is in studies of the microbial ecology of natural communities and the etiology of chronic bacterial infections that the synthesis of molecular and microscopic methods will produce an increment in understanding that will transform microbiology. This synthesis is already in place, in that we can examine material from natural and pathogenic ecosystems by confocal scanning laser microscopy (CSLM) and identify cells of particular taxonomic groups by the use of 16 S rRNA-specific FISH probes (Fig. 47). This techniques is especially valuable in well-studied ecosystems in which species diversity is relatively limited (e.g., single-species infections, see Fig. 61) or in which most of the community members have been grown in pure culture and identified (e.g., human vagina, see Fig. 41). In these molecular-based techniques for direct microscopy we can see and count individual bacterial cells, in relation to cells of other species and to host cells, and we can detect both bacterial and host reactions to these juxtapositions. Figure 47 shows a very active inflammatory process, with extensive mobilization of PMNs, so that we know that the location of these bacterial biofilms near tissues has elicited an exuberant phagocytic reaction. We can see the battlefield, we can identify the combatants, and we can finally stop trying to understand chronic infections by extrapolation from recovery-and-culture data and from in vitro studies of the recovered organisms in single-species culture.

Explosive new developments in population analysis by molecular techniques now offer us the golden opportunity of combining molecular and microscopic techniques for an even deeper understanding of the structure and function of natural and pathogenic communities. Roger Lasken's discovery of multiple displacement amplification (MDA), using a unique Phi 29 phage enzyme, now allows us to amplify the DNA from very small numbers (1 to 10) of prokaryotic or eukaryotic cells (Raghunathan et al. 2005) and to obtain sufficient DNA for full genomic characterization of these cells. Furthermore, mRNA can also be obtained from larger numbers (1200 to 1500) of prokaryotic or eukaryotic cells, and this pivotal nucleic acid can be used for quantitiative analyses of the expression of known genes by clonal aggregates of these cells. These are not projected estimates of what can be done in the future, but they are published data on the genomic analysis of single bacterial cells (Raghunathan et al. 2005) and gene expression data for clonal groups of ±1200 neural crest cells in "packets" migrating into the third brachial arch in mouse embryos (Bhattacherjee et al. 2004). The development of these new molecular techniques can now be linked to the equally explosive parallel development of new microscopic methods by the simple expedient of existing "capture" technologies that allow us to excise and recover cells for molecular analysis. Specifically, the new Zeiss 2 photon confocal microscope with PALM microdissection capability will allow us to visualize bacteria in natural and pathogenic ecosystems and then to recover clonal aggregates of either bacteria or eukaryotic cells for genomic or expressomic analysis. Once we know the 16 S rRNA sequence of an organism we have recovered by microdissection, we can check the database to determine whether it corresponds to any known species that has been grown in culture and we can then use FISH probes to locate it in samples of diseased tissues without ever culturing it.

We can now recline, in comfortable chairs with our favorite stimulants close at hand, and confidently imagine a new era of research in the study of bacterial communities in natural and engineered ecosystems and in the chronic diseases that vex and affect all of us. A united approach to microbial ecology will emerge, as direct observations linked with focused molecular analysis gradually replace extrapolation from in vitro studies of single-species cultures of organisms obtained by classic recovery-and-culture techniques. The new generation of microbial ecologists will link up across the anthrocentric divisions that fragment our science as they realize that the best way of knowing how many bacterial cells there are in a system is to stain and count them by direct microscopy. Acridine orange staining is not rocket science, and light microscopes are cheap, so we should no longer dismiss device-related infections as "aseptic loosening" when millions of bacteria are present and are easily detected by methods taught in Microbiology 101 labs. We will use 16 S RNA-directed FISH probes and fluorescent antibodies to identify bacterial cells of species that have already been isolated and characterized in vitro, and we will use the "optical sectioning" capabilities of the confocal microscope precisely to locate these cells in the three-dimensional context of their ecosystems. Figure 47 is a case in point in that we see biofilms formed by *Porphyromonas gingivalis* and "rosettes" formed by *Tanerella forsythensis* in their real spatial relationships to mobilized phagocytes and infected tissues in a real case of human periodontitis. We can use fluorescent antibodies to cytokines to study the inflammatory response of tissues and phagocytes to the juxtaposition of bacterial cells that have invaded their territory, and we can map the battlefield on which bacterial biofilms are challenged by host defenses.

If we ignore the siren call of infection and human calamity that so often pulls our science away from first principles, we can summarize the techniques that we can use to study bacteria directly and in situ in any ecosystem. We can use the high ribosome density of bacterial cells, and their consequent affinity for simple stains (e.g acridine orange, methylene blue), to locate all bacteria in a sample, and we can use confocal microscopy to visualize this sample in three dimensions. We can then use nonspecific FISH probes to locate all bacterial cells in broad categories (e.g., all Eubacteria or all Archeae) in their correct spatial relationships to physical surfaces and co-colonizing prokaryotes or host eukaryotic cells (Fig. 41). We can then use species-specific FISH probes and fluorescent antibodies to subdivide the broad bacterial categories and to identify and locate cells of known bacterial species (Figs. 47 and 61) in their correct spatial relationships to each other and to other morphological features of the ecosystem. We can identify the predominant biofilm organisms in any ecosystem, by DGGE (Fig. 62) and/or D-HPLC (Fig. 66), and we can then use FISH probes designed to react with their 16 S rRNA sequences (whether they have ever been cultured or not) to locate them precisely in the system or the infected tissue. Morphological keys are helpful in the construction of these detailed ecosystem maps, in that a large square-ended spore-forming rod that reacts with a bacillus-specific FISH probe can be tentatively identified as a bacillus species in subsequent examinations without further FISHing [sic!].

The inevitable questions concerning the viability and metabolic activity of the partners in microbial communities can now be addressed in situ by the use of the live/dead BacLight probe (Molecular Probes Part # L-7012) (Cook et al. 2000) and by the use of Micky Wagner's ingenious autoradiographic MAR technique (Daims et al. 2001). The BacLight probe simply reports the integrity of the bacterial cytoplasmic membrane in that propidium iodide penetrates to stain the DNA if the membrane is compromised and the cells stain red, while intact cells repel the propidium iodide and react with the counterstain (Syto 9) to stain green (Fig. 67). This probe must be used with

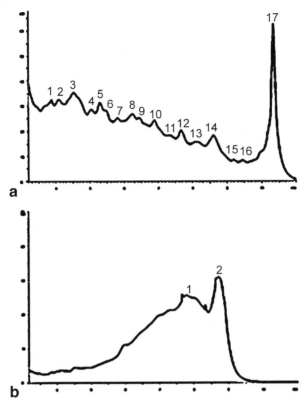

Fig. 66 Population analysis using HPLC to separate and recover 16 S rRNA genes from bacterial DNA extracted from human feces and amplified by PCR. **a** Presence of 17 distinct 16 S rRNA fractions in the feces of a human subject. Subsequent genomic analysis showed that only 5 of these peaks contained sequences that resembled those of any previously cultivated organisms. **b** D-HPLC pattern of DNA extracted from feces of same individual, following 21 d of oral ciprofloxacin and cotrimoxazole and intravenous vancomycin and meropenem. Note the radical change in species diversity to produce only 2 distinct peaks, both of which yielded sequences related to those of lactobacillus species. (From Goldenberg et al. 2006)

proper calibration, but it is useful in attesting to the general viability of mixed biofilm communities, and it especially useful in determining the extent to which antibacterial agents (e.g., sterilants and disinfectants) have killed bacterial cells in biofilms. The MAR technique (Daims et al. 2001) is potentially much more useful in that it enables us to determine the extent to which individual cells take up and metabolize specific radioactive substrates. Bacterial physiology has heretofore been predicated on studies of the average activity of millions of suspended planktonic cells, or of millions of cells growing in vitro in single-species biofilms, but we can now study chemical activities of individual cells in known locations in mixed-species communities.

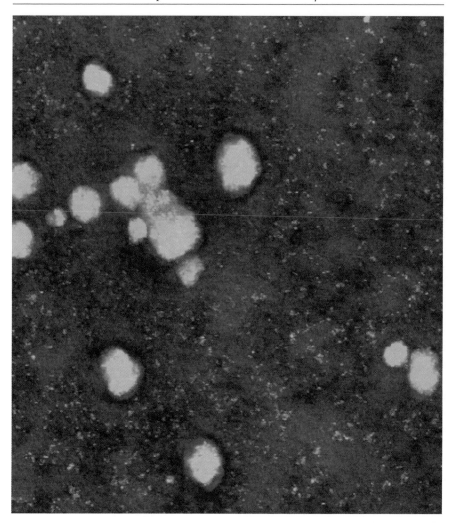

Fig. 67 Confocal micrograph, in x–y axis, showing killing of Phil Marsh's eight-species dental biofilm by an antibacterial solution. The treated biofilm was stained by the BacLite method, for the assessment of bacterial killing, and we note that the majority of the bacterial cells in the shallow biofilm are dead (*red*), while almost all of the cells in the raised "towers" have survived (*green*). Culture methods show a > 99% kill of this biofilm, by this agent, because the large aggregates do not grow when plated on agar

From our reservoir of mechanistic studies of bacterial physiology we know the "nuts and bolts" of most metabolic processes, and we know that the rates of these pivotal processes are controlled by a myriad of environmental factors, and now we have the opportunity to visualize these transformations in individual cells in communities. We can now examine the uptake and metabolism of a particular substrate molecule by individual cells of a particular species of bacteria, in immediate juxtaposition with cells of many different species or at various distances from cells of species that may produce enabling signal molecules. Or we can examine substrate processing by individual cells growing in a spectrum of oxygen concentrations, which we can determine in situ, and integrate this information to understand the obvious but mysterious effects of chemical gradients on bacterial activities in metabolically integrated communities. To this virtual cornucopia of in situ methods we can now add NMR microscopy (Majors et al. 2002), which allows us to obtain highly resolved NMR profiles of a huge variety of small molecules (e.g., organic acids) in small volumes (< 200 cubic μm) within biofilms that are simultaneously observed by confocal microscopy. We can determine the concentrations of these molecules continuously and in "real time" because the technique is nondestructive, so that we can feed a sugar to a biofilm and actually watch the production of acidic end products in a spatially resolved volume of a microbial community. Perhaps now we can rescue bacterial physiology from its shotgun marriage to bacterial genetics and revitalize the field to examine the humble premise that all cells of a given bacterial species do not conduct their metabolic business in identical ways in real multispecies ecosystems.

Perhaps the most exciting resolution of the most refractory problem in modern microbiology is our newly acquired ability to study microbial ecology without culturing all of the partners in the ecosystem to be examined. When we were trapped in the era of recovery-and-culture methodology, we became fixated on the species that could be cultured from what we knew to be mixed-species ecosystems of diabolical complexity. So *Streptococcus mutans* became our archetypical dental organism, and *E. coli* became our archetypical intestinal organism, without any concrete evidence for numerical predominance. Both organisms are important to human health, in oral and in digestive and urinary diseases, but modern methods of population analysis have shown that neither is present in predominant numbers (Fig. 66) in the ecosystem in which it was the putative ecological czar. Modern molecular methods of population analysis began with the extraction and PCR amplification of DNA from natural sources, and TIGR and the Venter Institute continue this laudable enterprise in ecosystems as diverse as the oceans and the human mouth. These data must be understood in terms of the samples used, and we must be prepared for seawater samples that happen to include a shred of kelp frond to differ radically from one that lack this garnish, but this approach will eventually yield a complete ecosystem census. Differential gradient gel electrophoresis (DGGE) yields very practical semiquantitative data on the predominant species present in an ecosystem (Fig. 62), and the refinement of this technique (D-HPLC), in which denatured DNA is separated on an HPLC column, has improved resolution (Fig. 66) and enabled us to recover the DNA for analysis using specific primers (Hurtle et al. 2002).

Two new technology streams now converge in that we can extract DNA from whole ecosystems and identify the predominant species present, and we can visualize bacterial cells in situ in actual ecosystems and amplify their DNA following precise microdissection of clonal aggregates. Both of these methods yield DNA from cells that have not been isolated, and may never have been cultured, but we know their entire genomes and we can determine their 16 S rRNA sequences and construct FISH probes to identify the individual cells in the mixed-species ecosystem. Of course, the process will be iterative, in that we will use D-HPLC to determine which species predominate in the whole ecosystem, and we will then use the FISH probes from that source to identify those bacterial cells in the intact ecosystem. If cells of particular interest (e.g., inflammation-causing cells) are then recovered by microdissection, we can then sequence their 16 S rRNA genes, construct the appropriate FISH probes, and ascertain whether cells within biofilms in inflamed tissues react with both probes. These technical developments will remove cultures from the equation, in that predominant organisms and organisms of special interest will be identifiable without ever having been isolated or cultured and their 16 S rRNA "signatures" can be added to the database that will serve all fields within microbiology.

How will these new methods and concepts affect microbiology? We have evolved from an approach in which we stood back from microbial communities that we judged to be too complex for comprehensive analysis, extracted a small fraction of the species present, and tried to explain community function in terms of these organisms of special interest. This approach worked well in the conquest of acute epidemic diseases and failed in the control of chronic biofilms diseases and in the analysis of natural and engineered ecosystems. In the new era, we will join the other biological sciences by using a combination of molecular and direct-observation techniques in an ambitious census of all of the bacterial species extant in all ecosystems. We will study the metabolic activities of individual cells, in situ in functioning ecosystems, and we will gradually come to understand the structure and function of the microbial communities that underpin the biosphere and undermine our health when we transgress against nature's laws!!

5 Toward a Unified Biofilm Theory

5.1
A Personal Odyssey

In the early 1980s, with the Scientific American publication (Costerton et al. 1978) and the alpine stream data foremost in my mind, I made a presentation to a bucolic Gordon Conference in which I suggested that *Pseudomonas aeruginosa* would adopt the biofilm mode of growth in infections of bladders, in burns, and in the lungs of cystic fibrosis patients. Harry Smith, who is the doyen of medical microbiology in the UK and in much of the known world, was in the audience, and I was concerned by his grim facial expression and horrified by his request (of the chairman) for an opportunity to rebut my talk in a special evening session. Harry was, and still is, my hero. During the free afternoon of that fateful day I walked round and round the one-quarter-mile track of the Tilton School trying to decide whether I would mount some form of defense against Harry's onslaught or whether I would simply roll over and concede defeat. I would have chosen the latter course, except for the one logical question I could not resolve, which was "How does Pseudomonas know where it is, in relation to the human body"? So the evening meal came and went, I presented myself for Harry's efficient and surgical evisceration with only a few desultory murmurs of argument, and we all carried away his message to the effect that it is stupid and sinful to extrapolate from one microbial system to another. Many years have now passed, and Harry traveled all the way to Canada with a lovely olive branch, but he raised a good question on which I will base the rest of this book: How much should we extrapolate from one organism to another or from one ecosystem to another?

When I went to Montana State University, to direct the Center for Biofilm Engineering (CBE), I was pole-axed by the sheer intelligence of engineers like Phil Stewart and Zbigniew Lewandowski, and I stumbled into another intellectual minefield that left me shaken and profoundly changed. With Anne Camper as my guide, since she speaks fluent "microbiology" and fluent "engineering", I slowly began to comprehend the "systems approach" to research and how engineers seek to establish general principles of system behavior with the objective of being able to predict system performance. I had always followed the scientific pattern of assembling an army of students and postdocs and running experiments and publishing papers until some kind of a pattern emerged, at which time I would propose some airy hypothesis and switch fields! The engineering systems approach is urgently needed in microbiology, and the time is ripe for its implementation.

We need to think about Woody Hasting's lab, with youngsters like Ken Nealson and Pete Greenberg, and Ned Ruby, and what might have happened if they had looked sideways to see if the genes for the Lux signaling system (Hastings and Nealson 1977) in *Vibrio harveyii* had homologues in *E. coli* and *P. aeruginosa*. Homologue identification was not really available in those days, but systems thinking would have saved 30 years of wandering in the wilderness and might have saved millions of lives. What would have happened if we had all looked for signs of coordinated behavior in our pet lab organisms, when Marty Dworkin and Dale Kaiser first described (Kaiser 1979; Dworkin 1983) swarming in myxobacteria, or if we had looked for exotic motilities and aggregation when Dale and Wenyuan Shi solved the mystery of gliding (Wall and Kaiser 1999; Shi and Zusman 1993) and identified the triggers for fruiting body formation? Microbiology has more silos than Kansas, and we have all watched the amazing commensal fluorescent marine bacteria, and the agile swarming cells of Listeria and subconsciously consigned these organisms to an "exotic" category far removed from our pet *E. coli* K 12 and *P. aeruginosa* PAO 1 that just sit around in test tubes and behave like Presbyterians. These are all bacteria, and the systems approach would insist that we accommodate their exotic behaviors in a general pattern that we need to establish if we are to predict how bacteria will perform in real ecosystems. So we would have looked for signal communication in Gram-negative bacteria in the 1960s, we would have looked for coordinated group mobility in the 1970s, and we would have looked for subtle invasions of host cells in the 1980s.

Perhaps because many of us are biochemists at heart, our first response to an "exotic" bacterial behavior is to construct mutants in which the behavior is modified or absent, and then to winkle out the gene and build a little mechanical model of the molecular machinery that drives the behavior. What we need to do now is to take these elegant little machines, especially the ones that accomplish various types of motility and chemotaxis, and use the tools of modern molecular biology and informatics to survey the Prokaryotic Kingdom and determine which organisms actually possess and use them. The new edition of Bergey's Manual is slimmer, but still pretty hefty, and we really don't need to add notations like "twitches" or "glides" to "ferments lactose but not fructose". There are "lumpers" and "splitters" in all intellectual communities, and we lump all bacteria together in their basic mechanisms of protein synthesis, and in their response to Bonnie Bassler's AI II signal (Xavier and Bassler 2003), but what else do the majority have in common? Do most of them swim, twitch, or glide? Do most of them respond to signals from other bacterial species or from eukaryotic hosts? Do most of them function as members of coordinated metabolic consortia? Does any significant number benefit from proximity to photosynthetic eukaryotes? What we do need is some cogent idea of what most bacteria do, in terms of metabolism and behavior when growing in biofilms in complex communities, and then we can be useful to engineers struggling with fouling and doctors trying to stem the tide of device-related and other chronic biofilm infections.

5.2
General Principles Underlying the Biofilm Theory

Nutrient sufficiency is the great watershed in microbial ecology, in that bacteria adopt their starvation strategies below certain levels of carbon and energy availability, and they form the ultramicrobacteria (Fig. 4) that are the very antithesis of biofilm communities. Above these threshold levels of nutrient availability, bacteria will adhere to surfaces and form biofilm communities (Figs. 1 and 11 and www.springer.com/978-3-540-68021-5: Movie 7), whose volume and extent are entirely dependent on the resources available to the community. While aggregates of cells that have detached from other biofilms (www.springer.com/978-3-540-68021-5: Movies 8 and 9) may be trapped in surface irregularities and begin to grow, the predominant cells involved in surface colonization are planktonic (www.springer.com/978-3-540-68021-5: Movie 7), and they show remarkably little discrimination in the types of surfaces to which they will adhere. Preferential colonization by individual species may occur if the incoming cells have enzymes that bind to surface materials (Fig. 25) or if tissue surfaces produce signals that attract cells of certain species (Figs. 37 and 40). Primary colonizers have been identified in many important biofilm systems, and many of these organisms have been shown to initiate mobile postadhesion behaviors during which they associate with cells of the same and metabolically cooperative species before forming aggregates. At a short time following initial adhesion, adherent cells begin to change their pattern of gene expression to their biofilm phenotype (Figs. 22 and 23), and their physical connection with the surface and to each other is altered by the secretion of polysaccharides and other matrix components (www.springer.com/978-3-540-68021-5: Movie 7). Surfaces may influence the resultant microbial communities, if they contain insoluble nutrients (e.g., cellulose, see Fig. 25) or reduced metal salts, because the biofilms will produce high local concentrations of enzymes and shuttle molecules to mobilize this energy. Similarly, plant and animal tissues may recruit cells of certain bacterial species by secreting signals that stimulate primary colonization, so that metabolically and/or ecologically cooperative biofilms may develop and be integrated into very efficient interkingdom communities (Figs. 18 and 39). The bona fides of these concepts are the remarkable microbial climax communities that dominate the biosphere in areas of major nutrient input (e.g., tidal flats and black smokers, see www.springer.com/978-3-540-68021-5: Movie 12) and occupy the surfaces of plant and animal tissues of pivotal importance to human life. The systems approach will be best served if we mobilize the powerful new techniques for direct observation and molecular analysis and study these biofilms, intact and in situ, until we can gradually understand the structure and function of the whole intact community.

As adherent bacteria adopt the biofilm phenotype, they retain their exquisite fine-tuned bacterial sensitivity to environmental stimuli, and each sessile bacterial cell in a biofilm will adjust its gene expression pattern to suit the precise microniche in which it finds itself (Costerton et al. 1994b). If conditions in the microniche change, the cell will change its phenotype, and it may even change its position in the community for better survival prospects. Signals from adjacent bacteria of cooperative and competing species will influence each sessile cell, to the same general extent as nutritional and/or environmental factors, and these communications may influence both position and activity (www.springer.com/978-3-540-68021-5: Movie 3). Individual cells in microbial communities that have "dedicated" ecological functions (e.g., methane genesis or cellulose digestion) will change their phenotypes until they find optimal positions and adopt optimal metabolic capabilities to serve the primary purpose of the community. The patterns of cell–cell association and of metabolic cooperativity (Fig. 1) that predominate in microbial biofilms with dedicated functions are "hard-wired" in the genomes of all bacterial species concerned, and they mobilize effective communities with striking regularity (Figs. 28 and 29). The location of individual bacterial cells in monospecies and mixed-species biofilms appears to be dictated by as yet undetermined species-specific rules, and we cannot be certain which classes of pili accomplish this positioning, but their location is certainly nonrandom (Fig. 10 and www.springer.com/978-3-540-68021-5: Movie 3). The boundaries of biofilms are open structures so that solutes can enter the communities unhindered, and the only elements that contribute homeostasis are the anionic components of certain matrix polymers (Fig. 32) that attract divalent cations (Ca^{++} and Mg^{++}), so that each sessile cell in a biofilm "sees" all of the molecules that diffuse through the community, including those in vesicles (Fig. 36) that may or may not be "addressed" to it, and responds to these stimuli to become a member of one or more functioning consortia. In communities that incorporate local areas of very high energy generation (e.g., reduced metal surfaces or photosynthetic neighbors), the matrix contains shuttle molecules and linear protein structures (nanowires) that transmit this energy more evenly throughout the biofilm. Biofilm communities also have the "option" of moving through their environment in swarming masses, coordinated by signals enclosed in vesicles, while retaining their spatial associations and metabolic integration and the capability of stopping to construct static communities in response to environmental cues. Integrated climax communities of truly remarkable efficiency exist in many ecosystems (e.g., bovine rumen and methanogenic granules), and we only need to study their processes in situ in order to plumb the outer limits of the metabolic efficiency of microbial consortia.

Multicellular communities of higher organisms are characterized by their ability to adapt to challenges and to respond to stimuli. We note that biofilms formed by cells of *P. aeruginosa* are more resistant if they have been exposed to antibiotics (Szomolay et al. 2005) or sterilants (Sanderson and Stewart 1997) in the recent past. Taken at face value, these observations suggest that biofilms can adapt to challenges and can communicate by signals both within and outside the contiguous community. We hereby invoke the systems approach to suggest that these communities can also respond to challenges, and send and receive signals, but we concede that the study of whole microbial communities is in its infancy and we cannot leap to conclusions. Biofilm cells up-regulate genes that generate diversity in microbial communities, and horizontal gene transfer is heavily favored by the stable juxtaposition of cells in these sessile communities (Fig. 10 and www.springer.com/978-3-540-68021-5: Movie 3), so that biofilms are inherently better equipped (Nguyen and Singh 2006) to withstand challenges by potentially lethal agents. The metabolic diversity of biofilms also favors community survival because cells of each component species grow in almost every conceivable metabolic state, somewhere in the community, and agents that attack fast-growing cells will fail to kill dormant cells. The survival of biofilms in chronic wounds provides the bona fides for this line of argument because these communities withstand the concerted challenge of dozens of antibiotics for years, in the absence of effective physical removal of the sessile bacteria by debridation. Cell–cell communication within biofilms is both documented and implicit in interspecies associations, but the outer limit of this activity may be in the conversations that allow mobile communities of myxobacteria to coordinate their activities, by vesicular signaling (Fig. 36), as they gallop through their soil habitat. We know the genes that allow squadrons of this microbial cavalry to settle down and build fruiting bodies when conditions deteriorate, and we are virtually obliged to see if homologous genes are involved in the production of Roberto Kolter's elegant arboreal spore "trees" by Bacillus species. The notion that microbial communities are sentient is so new that our minds race in circles, but the firm observation that linear protein nanowires run for hundreds of microns through these biofilms dangles mechanisms in front of our mind's eyes, and we are sorely tempted to speculate (www.springer.com/978-3-540-68021-5: Movie 3).

The remarkable tower, mushroom, and water channel structures (Figs. 1 and 11) we saw in in vitro grown biofilms of *P. aeruginosa* briefly set the standard of complexity in community architecture and served to inspire our biofilm modelers to declare this to be the optimal arrangement (www.springer.com/978-3-540-68021-5: Movie 10) for solute exchange with the bulk fluid. But the outer milepost in cellular architecture was soon usurped by Staffan Kjelleberg's towers, bridges, and rosettes formed by *Serratia liquefaciens* (Labbate et al. 2004) and Tim Tolker-Nielsen's stump and mushroom cap architecture in plain old *P. aeruginosa* (PAO 1) (Klausen et al. 2003). In the systems approach, we can distill these observations to state that cells are precisely positioned within biofilms and that cells can move from one configuration to another, which presumes both a positioning mechanism and one or more translation mechanisms. As John Lawrence begins to explore real environmental biofilms and Christoph Schaudinn rappels into the gingival crevice, the first images (Fig. 13) they send back to home base simply take our breath away. There seem to be no limits on the variety of cellular arrangements in real biofilms, and the cellular architecture does not always correspond exactly to the matrix architecture, which suggests that they may be independent. The matrices of several microbial communities have been shown to transcend our simple enveloping slimy polysaccharide model, in that a semirigid scaffolding is constructed on which bacterial cells are deployed in positions (Fig. 10) that favor oxygen circulation and enhanced metabolic activity. The honeycomb "apartment blocks" made by the MH strain of *S. epidermidis* provide another milepost, in our systems approach, in that prokaryotic cells in a biofilm community have been shown (Figs. 14 to 17) to be able to synthesize and organize complex structures of a size and complexity that taxes our imaginations. Microbial communities can position their component cells, move these cells to produce defined patterns, synthesize and position a variety of matrix components to create microcolonies and water channels, and synthesize and organize very complex tertiary structures in the intercellular space. If Harry Smith will let us generalize from these few salient examples, the differences between microbial biofilm communities and eukaryotic tissues begin to blur, and bacteria move into a much higher position in the scheme of living things.

5.3
The Biofilm Theory Can Unite and Revitalize Microbiology

American universities are closing and amalgamating microbiology departments at an alarming and accelerating rate. Those departments that buck this trend survive by finding a molecular focus (e.g., cell–cell signaling) that funds major research projects and small armies of graduate students and postdocs, while dwindling undergraduate numbers cause apoplexy in the provost's office. Meanwhile, dozens of microbiologists labor away in other campus units in which our anthrocentric traditions have placed them. They will be squirreled away in environmental engineering, in food soil science, and in the departments within medicine and dentistry that deal with infectious processes, and a small contingent will doze peacefully in oenology!

Microbiologists can recapture the academic and scientific initiative, and the high ground of academic politics, by adding back the study of microbial communities to the course offerings of their microbiology departments. Processes of intense interest to environmental engineers, agriculturalists, doctors, and dentists have the common denominator of being carried out by organized microbial communities. If we replace our moribund courses in microbial physiology with new and innovative courses in the structure and function of microbial communities, we can reverse the microbiological diaspora on our campuses and pack our lecture halls. If we can interest the folks with the big grants and shiny equipment in using their powerful molecular techniques in the analysis of natural and pathogenic microbial communities, then we can convert our hospital labs from culture methods to PCR DGGE and D-HPLC. Senior courses in population analysis by molecular techniques would unite the scattered microbiologists in engineering, agriculture, and medicine because the same methods would be used to study microbial communities in wastewater, soil, and refractory infections. And if we convert our dreary Microbiology 101 courses from species catalogs with plate-streaking labs to microbial ecology courses with FISH probe microscopy and molecular population analysis, we will need to build new classrooms!

Our peculiar history of serving humankind by raising cadres of microbiologists to address and solve critical problems as they arose has produced a compartmentalized community and an ingrained insularity of thought processes. Instead of thinking broadly, like a chemist mixing an acid with an ester (bad idea), we have created pigeonholes in which we place bacterial species and memorize and marvel at their properties. The study of microbial communities will have a salutary effect on microbiology because we will identify bacterial species by their roles (e.g., primary colonizers, nitrite oxidizers), and their behavior in pure liquid cultures will recede into insignificance. As we take a more general view of our small subjects, we will look for similarities in species with the same roles and characteristics when we find a mechanism or behavior in a particular species. If we find a phenomenon like quorum-based signaling in a particular species, we will not wait 20 years to search for the same capability in other species. If we see bacteria of a certain species moving through their ecosystems in coordinated swarms, we will grab a microscope and look at other ecosystems to see who might be doing the same, and we will identify them by means of FISH probes. We will always use the species concept, because millions of genomically distinct organisms make up the microbial world, but the concept of coordinated communities with different species involved in identical roles (e.g., nitrate reduction) will induce us to abandon our pigeonholes and think globally. Perhaps when the similarities of metabolically coordinated communities are recognized across a broad range of ecosystems and when similar molecular mechanisms are found in the species that comprise these communities, microbiologists will unite in a revitalized community of our own.

5.4
The Biofilm Theory

5.4.1
Narrative

Direct observations of bacteria, in natural and pathogenic ecosystems, have shown that more than 99% of these organisms grow and function as members of metabolically integrated communities (Fig. 1). For this reason, the genome of each individual cell and of its metabolically integrated companions will have been shaped by selection pressures that favor their integration into this particular community and into other communities of which they may be equally valuable members. If we consider the behavior of an individual planktonic cell as a starting point in this narrative, we can muster some sympathy as the newly independent cell detaches from its crowded community and launches itself into the bulk fluid. Many hazards await, its chances of finding another home are very slim indeed, and it faces a prolonged period in starvation/oblivion in the stygian depths of the Earth or ocean if it survives all of the threats to which planktonic bacterial cells are very susceptible.

The planktonic cell is, however, equipped with receptors tuned to the signals and nutrients it experienced in its halcyon days in its home community. It can move up-gradient toward these familiar signals and nutrients, and this ability may bring it into contact with a new community in the act of forming in a favorable location. The signals to which it is attracted may originate from a living tissue, or they may be produced by primary colonizers that have preceded it in the mobilization of the community, and some nutrients may seem like Mother's apple pie to the lonely tyke. If our hypothetical cell has no receptors for the signals, or if it has never had apple pie, it will continue toward oblivion, but it has the opportunity to join the forming community if its genome was shaped by life in its home community and it recognizes familiar signals and molecules. Even if the forming community is not entirely familiar, our cell may recognize some signals and may join the scrum of other planktonic cells in transient associations with each other and with the available surface. Even if our hypothetical cell does not find a receptive forming community, it may find an empty surface and become a primary colonizer with the potential for attracting cooperative species by its own distinct signal pattern. Once the cell has joined the nascent community, its association with cells of other species on the colonized surface will depend on its production of the "correct" signals and structures that allow it to "plug in" to the cell–cell communication and energy distribution systems of the community. If the genetic hard-wiring of the cell matches that of the community, it will become a productive member of one or more metabolic consortia, but it will languish like a lost soul if it has blundered into a community into which it does not fit. Mature biofilm communities produce and detach billions of planktonic cells, individual cell life is "cheap", and the community is the unit of survival that matters in the natural world.

Once a cell has found a community in which the functions and associations dictated by its genome are well matched by those of the other members, a large number of attractive lifestyles beckon. Specialized community members may obtain energy from surface reactions, like iron oxidation or cellulose digestion, and may disseminate this energy throughout the whole community by means of nanowires or redox couples. Consortia of several species may digest complex nutrients, like bitumen and leaf litter, and the protons they produce in their exuberance will be trapped in the biofilm matrix and shared with other community members. If the main metabolic activities of the community demand oxygen or other diffusible nutrients, the metabolic units will form distinct separate microcolonies separated by open water channels through which the vital nutrients can diffuse and wastes can be removed. The metabolic activities of the biofilm community may be integrated with those of a living eukaryotic animal or tissue, and the bacterial community may serve its host by physiological contributions (e.g., urea reduction) or by protection against bacterial pathogens. These cell associations and overall biofilm architectures depend on cell–cell connections, such as pili and nanowires, and this network facilitates horizontal gene transfer and positions cells in what constitutes a self-assembled multicellular organism.

Metabolically integrated bacterial communities often continue their operations for centuries, where tidal flats or hot springs supply habitat and favorable conditions on a continual basis, but most biofilms live less idyllic lives. Many biofilms grow on nutrient surfaces (e.g., plant material) that are consumed by the community activity, so that its members must detach and reassemble on fresh surfaces, and many biofilms stagnate when their end products are not removed by flow. While these shifting nutrient strategies lead to the detachment of individual planktonic cells, whole bacterial communities may adopt the more effective strategy of moving through their ecosystem in coordinated swarms to reach a whole series of fresh nutrient sources. These microbial communities are not limited by rigid boundaries, and spatially connected consortia that accumulate inhibitory levels of their own end products may have these molecules removed by mobile allies that are never structurally integrated into the community. Nutrients are limiting in most microbial ecosystems, so the communities that can fix and process these resources most efficiently will thrive and be perpetuated, and they can adapt their membership patterns and their architecture in any manner that promotes this efficiency.

A bacterial biofilm presents an attractive target to presumptuous biological predators, from bacteriophages to amoebae, and the continuing predominance of these communities in the entire biosphere constitutes a stern rebuke to this insolence. Viruses usually cannot reach their targets in the cell envelopes of bacterial cells encased in matrix materials, and mixed-species biofilms in real ecosystems contain millions of these frustrated molecular missiles in tangled suspension. Direct observations of natural river biofilms are vastly entertaining for biofilm fans because hulking amoebae can be seen to capture and digest single bacterial cells, while intact biofilms repel their overtures completely. Small biofilm microcolonies are phagocytized and expelled without any apparent damage to their component cells, and free-living amoebae really seem to be gardeners in vast biofilm jungles, trimming and pruning but never consuming the main crop. Snails and insect larvae rasp and tunnel in the thick mature biofilms of all eutrophic aquatic systems, but the microbial population quickly reasserts itself, and Roberto Kolter's aquarium bears silent witness to the eventual triumph of the prokaryotes. Some predators and competitors attack biofilms with untargeted chemical agents, but these communities adapt to these attacks by constructing their matrices out of materials that bind the most common aggressive molecules. When these predators use targeted antibacterial agents (antibiotics), they usually fail because the biofilm cells have invoked special recombination genes to generate sufficient diversity so that some members will always have modified their target molecules and will persist.

The ecosystems that present the greatest challenge to bacterial communities, with their variable membership and their hard-wired strategies, are those of the human body and of our most valued domestic animals and plant crops. This is because we can devise strategies of our own, and we can counteract the bacterial strategies if we understand them. Animal ecosystems have always been inherently forbidding to bacteria because of the fine-tuned antibodies and phagocytes of the adaptive immune system and because of the barriers erected in each organ system by anatomy, physiology, or commensal bacterial populations. Bacteria have countered by lurking in biofilms in natural ecosystems in the neighborhood and "waiting" until large human populations lost their adaptive immunity and relaxed their bactericidal and opsonizing immune defenses. These specialized pathogens would then emerge from hiding and use some universal vector (e.g., water supply or parasitic insects) to mount acute planktonic attacks that killed humans before they could mount an effective immune response. Humans then countered by stimulating preemptive immune responses with vaccines and by controlling bacterial transmission by water and insects, and true epidemics became a thing of the past in the developed world. Specialized pathogens then hunkered down in ecosystems close to compromised humans and mounted regular acute planktonic attacks on their victims that were successful, until we cleaned up our hospitals and developed antibiotics to kill the planktonic attackers. Human deaths from bacterial disease were reduced to a certain level, as early as the 1960s and 1970s, but we have been unable to reduce them to nil because of the inevitable bacterial counterattack based, this time, on a dual strategy. The bacteria have mutated to forms that either lack the specific targets of conventional antibiotics or to forms that can protect these targets, so that they can continue to lurk in hospital environments and mount frontal planktonic attacks on compromised patients. And bacteria have expanded their oldest and most effective strategy by extending the biofilms formed by ordinary environmental organisms onto the surfaces of inert biomaterials and of compromised tissues and taking a passive/aggressive stance of persisting and inviting the human counterattack. Perhaps this second and most common strategy is the most successful to date because the bacteria benefit from an optimal environment for much longer periods, and they do not need toxins because we humans damage ourselves in our frenzied inflammatory response to their continuing presence. The next move is ours, now that we understand the game, and it will almost certainly involve specific inhibition of biofilm formation and modulation of the inflammatory response.

5.4.2
Summary

Bacteria live preferentially in multicellular biofilms in which cells are positioned for optimal metabolic interaction, and the resultant architecture favors the ecological role of the community in the ecosystem. These biofilm communities have developed structures and strategies in response to attacks by chemical and biological antagonists, and successful architectural and defensive strategies are both "hard-wired" into the metagenomes of all species. The biofilm is the basic evolutionary unit that repeatedly self-assembles from multiple genomes, in ecosystems in which it is successful, and these communities detach large numbers of planktonic cells for the downstream dissemination of these genomes. The planktonic phenotype differs radically from the much more numerous biofilm phenotype, and this shift in gene expression favors rapid growth and mobility but makes the individual cells much more susceptible to antibacterial agents. Bacterial depredations that affect humans initially involved the planktonic phenotype, in the form of acute infections, but human countermeasures have largely contained this threat and it has been largely replaced by device-related and other chronic biofilm infections.

5.4.3
Definition

A biofilm is a multicellular community composed of prokaryotic and/or eukaryotic cells embedded in a matrix composed, at least partially, of material synthesized by the sessile cells in the community.

5.5
The Way Forward

Two observations must stand until they are contradicted by new data:

1. The majority of bacteria in natural ecosystems, and in chronic infections, can be seen to grow in biofilms.
2. Bacteria growing in biofilms adopt a phenotype that is significantly different from that of their planktonic counterparts.

To the researchers among us this means that we should examine our system of interest in both phenotypes. We can apply concepts derived from the study of planktonic cells only in ecosystems in which the organism grows predominantly as planktonic cells, and we should apply concepts derived from the study of the biofilm phenotype to systems in which biofilms predominate. We must acknowledge that subculture often selects mutants whose genome differs profoundly from the metagenome of wild strains of the species concerned, and we should recognize the importance of environmental factors by testing our conclusions in real ecosystems as soon as possible.

To the clinicians among us this means that we should recognize data that are derived from planktonic cells, such as antibiotic susceptibility values, can only apply to the acute phase of the infection being treated. We should acknowledge that fragments of biofilms that are recovered from colonized or infected tissues will not grow to produce visible colonies on agar media, and we should interpret culture data only as they reflect the presence and susceptibility of planktonic cells. We should encourage the adoption of nucleic-acid-based methods for the detection of pathogens growing in the biofilm phenotype and use commercial biofilm inhibitors as soon as they are FDA-approved and available. We should acknowledge that continued bacterial persistence and inappropriate host responses cause much of the damage in biofilm infections, and we should consider the use of suitable agents to manipulate this inflammatory response

References

Adair CG, Gorman SP, Feron BM, Byers LM, Jones DS, Goldsmith CE, Moore JE, Kerr JR, Curran MD, Hogg G, Webb CH, McCarthy GJ, Milligan KR (2004) Implications of endotracheal tube biofilm for ventilator-associated pneumonia. Intensive Care Med 25:1072–1076

Adam RD (2001) Biology of *Giardia lamblia*. Clin Microbiol Rev 14:447–475

Allegrucci M, Hu FZ, Shen K, Hayes J, Ehrlich GD, Post JC, Sauer K (2006) Phenotypic characterization of *Streptococcus pneumoniae* biofilm development. J Bacteriol 188:2325–2335

Amann RI, Ludwig W, Schleifer KH (1995) Phylogenetic identification and in situ detection of individual microbial cells without cultivation. Microbiol Rev 25:143–169

Anderson GG, Palermo JJ, Schilling JD, Roth R, Heuser J, Hultgren SJ (2003) Intracellular bacterial biofilm-like pods in urinary tract infections. Science 301:105–107

Auerbach HS, Williams M, Kirkpatrick JA, Colten HR (1985) Alternate day prednisone reduces morbidity and improves pulmonary function in cystic fibrosis. Lancet 292:686–688

Balaban N, Goldkorn T, Nhan RT, Dang LB, Scott S, Ridgley RM, Rasooly A, Wright SC, Larrick JW, Rasooly R, Carlson JR (1998) Autoinducer of virulence as a target for vaccine and therapy against *Staphylococcus aureus*. Science 280:438–440

Balaban N, Giacometti A, Cirioni O, Gov Y, Ghiselli R, Mocchegiani F, Viticchi C, Del Prete MS, Saba V, Scalise G, Dell'Acqua G (2003) Use of the quorum-sensing inhibitor RNAIII-Inhibiting peptide to prevent biofilm formation in vivo by drug-resistant *Staphylococcus epidermidis*. J Infect Dis 187:625–630

Balkwill D, Reeves RH, Drake GR, Reeves JT, Crocker, FH, King MB, Boone DR (1997) Phylogenetic characterization of bacteria in the subsurface microbial culture collection. FEMS Microbiol Rev 20:201–219

Bell CR, Holder-Franklin MA, Franklin M (1982) Correlations between predominant heterotrophic bacteria and physicochemical water quality parameters in two Canadian rivers. Appl Environ Microbiol 43:269–283

Beloin C, Valle J, Latour-Lambert P, Faure P, Kzieminski P, Balestrino D, Haagensen JAJ, Molin S, Prensier G, Arbeille B, Ghigo J-M (2004) Global impact of mature biofilm lifestyle on Escherichia coli K-12 gene expression. Mol Microbiol 51:659–665

Beveridge TJ (2006) Visualizing bacterial cell walls and biofilms. Microbe 1:279–284

Bhattacherjee VP, Mukhopadhyay P, Singh S, Roberts EA, Hackmiller RC, Greene RM, Pisano MM (2004) Laser capture microdissection of fluorescently labeled embryonic cranial neural crest cells. Genesis 39:58–64

Blenkinsopp SA, Khoury AE, Costerton JW (1992) Electrical enhancement of biocide efficacy against *Pseudomonas aeruginosa* biofilms. Appl Environ Microbiol 58:3770–3773

Bockelmann U, Janke A, Kuhn R, Neu TR, Wecke J, Lawrence JR, Szewzyk U (2006) Bacterial extracellular DNA forming a defined network-like structure. FEMS Microbiol Lett 262:31–38

Bodaly RA, Hecky RE, Fudge RJP (1984) Increases in fish mercury levels in lakes flooded by the Churchill River diversion. Can J Fish Aquat Sci 41:682–691

Boles BR, Thoendel M, Singh PK (2004) Self-generated diversity produces "insurance effects" in biofilm communities. Proc Natl Acad Sci USA 101:16630–16635

Boyd A, Chakrabarty AM (1994) Role of alginate lyase in cell detachment of *Pseudomonas aeruginosa*. Appl Environ Microbiol 60:2355–2359

Brady RA, Leid JG, Camper AK, Costerton JW, Shirtliff ME (2006) Identification of *Staphylococcus aureus* proteins recognized by the antibody-mediated immune response to a biofilm infection. Infect Immun 74:3415–3426

Bresnak JA, Brune A (1994) Role of microorganisms in the digestion of lignocellulose by termites. Annu Rev Entomol 39:453–487

Burr MD, Clark SJ, Spear CR, Camper AK (2006) Denaturing gradient gel electrophoresis can rapidly display the bacterial diversity contained in 16 S rDNA clone libraries. Microb Ecol 51:479–486

Caccavo JR, Blakemore RP, Lovely DR (1992) A hydrogen-oxidizing, Fe III reducing microorganism from the Great Bay estuary, New Hampshire. Appl Environ Microbiol 58:3211–3216

Caldwell DE, Costerton JW (1996) Are bacterial biofilms constrained to Darwin's concept of evolution through natural selection? Microbiologica 12:347–358

Caldwell MB, Walker RI, Stewart SD, Rogers JE (1983) Simple adult rabbit model for *Campylobacter jejuni* enteritis. Infect Immun 42:1176–1182

Ceri H, Olson ME, Stremick C, Read RR, Morck D, Buret A (1999) The Calgary Biofilm Device: New technology for rapid determination of antibiotic susceptibilities of bacterial biofilms. J Clin Microbiol 37:1771–1776

Chan RCY, Reid G, Bruce AW, Costerton JW (1984) Microbial colonization of human ileal conduits. Appl Environ Microbiol 48:1159–1165

Cheng K-J, Fay JP, Howarth R, Costerton JW (1980) Sequence of events in the digestion of fresh legume leaves by rumen bacteria. Appl Environ Microbiol 40:613–625

Cheng K-J, Costerton JW (1981) Adherent rumen bacteria: their role in the digestion of plant material, urea, and epithelial cells. In: Ruchebusch Y, Thivend P (eds) Digestive Physiology and Metabolism in Ruminants. MTP Press, Lancaster, UK, pp 227–250

Cheng K-J, Costerton JW (1986) Benefiting ruminants by manipulation of bacteria that degrade fibrous feeds and adhere to the digestive tract surfaces. In: Waldvogel E (ed) Ruminant Digestion. Elsevier, Amsterdam, pp 121–127

Christner BC, Mosley-Thompson E, Thompson LG, Reeve JN (2001) Isolation of bacteria and 16 S rDNAs from Lake Vostok accretion ice. Environ Microbiol 3:570–578

Cochrane DMG, Brown MRW, Anwar H, Weller PH, Lam K, Costerton JW (1988) Antibody response to *Pseudomonas aeruginosa* surface protein antigens in a rat model of chronic lung infection. J Med Microbiol 27:255–261

Colwell R, Huq A (2001) Marine ecosystems and cholera. Hydrobiologia 460:141–145

Cook G, Costerton JW, Darouiche RO (2000) Direct confocal microscopy studies of the bacterial colonization in vitro of a silver-coated heart valve sewing cuff. Int J Antimicrob Agents 13:169–173

Costerton JW, Stewart PS, Greenberg EP (1999) Bacterial biofilms: a common cause of persistent infections. Science 284:1318–1322

Costerton JW (2004) Microbial ecology comes of age and joins the general ecological community. Proc Natl Acad Sci USA 101:16983–16984

Costerton JW (2005) Biofilm theory can guide the treatment of device-related orthopedic infections. Clin Orthop Rel Res 437:7–11

Costerton JW, Lappin-Scott HM (1995) Introduction to microbial biofilms. In: Lappin-Scott HM, Costerton JW (eds) Microbial Biofilms. Cambridge University Press, Cambridge, UK, pp 1–11

Costerton JW, Stewart PS (2001) Battling Biofilms. Sci Am 285:75–81

Costerton JW, Geesey GG, Cheng K-J (1978) How bacteria stick. Sci Am 238:86–95

Costerton JW, Cheng K-J, Geesey GG, Ladd TI, Nickel JC, Dasgupta M, Marrie TJ (1987) Bacterial biofilms in nature and disease. Annu Rev Microbiol 41:435–464

Costerton JW, Ellis B, Lam K, Johnson F, Khoury AE (1994a) Mechanism of electrical enhancement of efficacy of antibiotics in killing biofilm bacteria. Antimicrob Agents Chemother 38:2803–2809

Costerton JW, Lewandowski Z, deBeer D, Caldwell D, Korber D, James G (1994b) Minireview: biofilms, the customized micronich. J Bacteriol 176:2137–2142

Costerton JW, Lewandowski Z, Caldwell DE, Korber DR, Lappin-Scott HM (1995) Microbial biofilms. Annu Rev Microbiol 49:711–745

Costerton JW, Stewart PS, Greenberg EP (1999) Bacterial biofilms: a common cause of persistent infections. Science 284:1318–1322

Costerton JW, Veeh R, Shirtliff M, Pasmore M, Post C, Ehrlich G (2003) The application of biofilm science to the study and control of chronic bacterial infections. J Clin Invest 112:1466–1477

Crawford DL, Crawford RL, Pometto AL (1977) Preparation of specific labeled ^{14}C-(Lignin)- and ^{14}C-(Cellulose)-Lignocelluloses and their decomposition by the microflora of soil. Appl Environ Microbiol 33:1247–1251

Cusack FM, Lappin-Scott HM, Singh S, de Rocco M, Costerton, JW (1990) Advances in microbiology to enhance oil recovery. Appl Biochem Biotechnol 24/25:885–898

Cusack F, Singh S, McCarthy C, Grieco J, de Rocco M, Nguyen D, Lappin-Scott H, Costerton JW (1992) Enhanced oil recovery: three dimensional sandpack simulation of ultramicrobacteria resuscitation in reservoir formations. J Gen Microbiol 138:647–655

Daims H, Nielsen JL, Nielsen PH, Schleifer K-H, Wagner M (2001) In situ characterization of Nitrospira-like nitrite-oxidizing bacteria active in wastewater treatment plants. Appl Environ Microbiol 67:5273–5284

Dasgupta MK, Bettcher KB, Ulan RA, Burns V, Lam K, Dossetor JB, Costerton JW (1987) Relationship of adherent bacterial biofilms to peritonitis in chronic ambulatory peritoneal dialysis. Peritoneal Dialysis Bull 7:168–173

Davey ME, Costerton JW (2006) Molecular genetics analyses of biofilm formation in oral isolates. Periodontology 42:1–14

Davies DG, Geesey GG (1995) Regulation of the alginate biosynthesis gene algC in Pseudomonas aeruginosa during biofilm development in continuous culture. Appl Environ Microbiol 61:860–867

Davies DG, Parsek MR, Pearson JP, Iglewski BH, Costerton JW, Greenberg EP (1998) The involvement of cell-to-cell signals in the development of a bacterial biofilm. Science 280:295–298

deBeer D, Stoodley P, Roe F, Lewandowski Z (1994) Effects of biofilm structures on oxygen distribution and mass transport. Biotechnol Bioeng 43:1131–1138

De Kievit TR, Gillis R, Marx S, Brown C, Iglewski BH (2001) Quorum-sensing genes in Pseudomonas aeruginosa biofilms: their role and expression patterns. Appl Environ Microbiol 67:1865–1873 de Kruif P (1926) The Microbe Hunters. Harcourt Brace, New York

de Nys R, Steinberg PD, Willemsen P, Dworjanyn SA, Gabelish CL, King RJ (1995) Broad spectrum effects of secondary metabolites from the red alga *Delisea pulchra* in antifouling assays. Biofouling 8:259–271

Dohar JE, Hebda PA, Veeh R, Awad M, Costerton JW, Hayes J, Ehrlich GD (2005) Mucosal biofilm formation on middle-ear mucosa in a nonhuman primate model of chronic suppurative otitis media. Laryngoscope 115:1469–1472

Donlan RM (2001) Biofilm formation: a clinically relevant microbiological process. Clin Infect Dis 33:1387–1392

Donlan RM, Costerton JW (2002) Bioflms: survival mechanisms of clinically relevant microorganisms. Clin Microbiol Rev 15:167–193

Doring G, Conway SP, Heijerman HG, Hodson ME, Hoiby N, Smyth A, Touw DJ (2000) Antibiotic therapy against *Pseudomonas aeruginosa* in cystic fibrosis: a European consensus. Eur Respir J 16:749–767

Douglas LJ (2003) Candida biofilms and their role in infection. Trends Microbiol 11:30–36

Drenkard E, Ausubel FM (2002) *Pseudomonas* biofilm formation and antibiotic resistance are linked to phenotypic variation. Nature 416:740–743

Dunny GM, Leonard BA (1997) Cell-cell communication in Gram-positive bacteria. Annu Rev Microbiol 51:527–564

Dupraz C, Visscher PT, Baumgartner LK, Reid RP (2004) Microbe-mineral interactions: early carbonate precipitation in a hypersaline lake (Eleuthera Island, Bahamas). Sedimentology 51:745–776

Dutta L, Nuttall HE, Cunningham A, James G, Hiebert R (2005) In situ biofilm barriers: case study of a nitrate groundwater plume, Albuquerque, New Mexico. Remediation 15:101–111

Dworkin M (1983) Tactic behavior of *Myxococcus xanthus*. J Bacteriol 154:452–459

Ehrlich GD, Hu FZ, Shen K, Stoodley P, Post JC (2005) Bacterial plurality as a general mechanism driving persistence in chronic infections. Clin Orthop Rel Res 437:20–24

Espinoza J, Erez O, Romero R (2006) Preconceptional antibiotic treatment to prevent preterm birth in women with a previous preterm delivery. Am J Obstetr Gynecol 194:630–637

Feldman GL, Krezanoski JZ, Ellis BD, Lam K, Costerton JW (1992) Control of bacterial biofilms on rigid gas permeable lenses. Spectrum 1992:36–39

Fletcher M (1987) How do bacteria attach to solid surfaces? Microbiol Sci 4:133–136

Fuqua WC, Greenberg EP (2002) Listening in on bacteria: acyl-homoserine lactone signaling. Nat Rev Mol Cell Biol 3:685–695

Fuqua WC, Winans EP, Greenberg EP (1994) Quorum sensing in bacteria: the Lux R - Lux I family of cell density-responsive transcriptional regulators. J Bacteriol 176:269–275

Fux CA, Stoodley P, Hall-Stoodley L, Costerton JW (2003) Bacterial biofilms: a diagnostic and therapeutic challenge. Expert Rev Anti-Infect Ther 1:667–683

Fux CA, Wilson S, Stoodley P (2004) Detachment characteristics and oxacillin resistance of *Staphylococcus aureus* biofilm emboli in an in vitro catheter infection model. J Bacteriol 186:4486–4491

Fux CA, Costerton JW, Stewart PS, Stoodley P (2005a) Survival strategies of infectious biofilms. Trends Microbiol 13:34–40

Fux CA, Shirtliff M, Stoodley P, Costerton JW (2005b) Can laboratory reference strains mirror real-world pathogenesis? Trends Microbiol 13:58–63

Gantner S, Schmid MA, Durr C, Schuhegger R, Steidle A, Huntzler P, Langebartels C, Eberl L, Hartmann A, Dazzo FB (2006) In situ quantitation of the spatial scale of calling distances and population density-independent N-acylhomoserine lactone-mediated communication by rhizobacteria colonized on plant roots. FEMS Microbiol Ecol 56:188–194

Geesey GG, Richardson WT, Yeomans HG, Irvin RT, Costerton JW (1977) Microscopic examination of natural sessile bacterial populations from an alpine stream. Can J Microbiol 23:1733–1736

Geis G, Leying H, Suerbaum S, Mai U, Opferkuch W (1989) Ultrastructure and chemical analysis of *Campylobacter pylori* flagella. J Clin Microbiol 27:436–441

Ghannoum M, O'Toole GA (2004) Microbial Biofilms. ASM Press, Washington, DC, p 426

Ghigo J-M (2001) Natural conjugative plasmids induce biofilm development. Nature 412:442–445

Gibbons RJ, van Houte J (1975) Dental caries. Annu Rev Med 26:121–136

Gilbert P, Maira-Litran T, McBain AJ, Rickard AH, Whyte FW (2002) The physiology and collective recalcitrance of microbial biofilm communities. Adv Microbiol Physiol 46:2002–2056

Goldenberg O, Herrmann S, Marjoram G, Noyer-Weidner M, Hong G, Bereswill S, Gobel UB (2006) Molecular monitoring of the intestinal flora by denaturing high performance liquid chromatography. J Microbiol Methods (Epub ahead of print)

Gorby YA, Yanina S, McLean JS, Russo KM, Moyles D, Dohnalkova A, Beveridge TJ, Chang IS, Kim BH, Kim KS, Culley DE, Reed SB, Romine MF, Saffarini DA, Hill EA, Shi L, Elias DA, Kennedy DW, Pinchuck G, Watanabe K, Iishi SI, Logan B, Nealson KH, Fredrickson JK (2006) Electrically conductive bacterial nanowires produced by Shewanella oneidensis strain MR-1 and other microorganisms. Proc Natl Acad Sci USA 103:11358–11363

Greene K (2002) New method for culturing bacteria. Science 296:1000

Grimes J (2006) Koch's postulates – then and now. Microbe 1:223–228

Gristina AG, Costerton JW (1984) Bacteria-laden biofilms: a hazard to orthopedic prostheses. Infect Surg 3:655–662

Gristina AG, Hobgood CD, Webb LX, Myrvik QN (1987) Adhesive colonization of biomaterials and antibiotic resistance. Biomaterials 8:423–426

Gristina AG, Dobbins JJ, Giamara B, Lewis JC, DeVries WC (1988) Biomaterial-centered sepsis and the total artificial heart: microbial adhesion versus tissue integration. J Am Med Assoc 259:870–877

Hall-Stoodley L, Costerton JW, Stoodley P (2004) Bacterial biofilms: from the natural environment to infectious diseases. Nat Rev Microbiol 2:95–108

Hall-Stoodley L, Hu FZ, Gieseke A, Nistico L, Nguyen D, Hayes J, Forbes M, Greenberg DP, Dice B, Burrows A, Wackym PA, Stoodley P, Post JC, Ehrlich GD, Kerschner JE (2006) Direct detection of bacterial biofilms on the middle ear mucosa of children with otitis media. J Am Med Assoc 296:202–211

Hamilton M, Johnson K, Camper A, Stoodley P, Harkin G, Gillis R, Shope P (1995) Analysis of bacterial spatial patterns at the initial stage of biofilm formation. Biometr J 37:393–408

Harford CG, Leidler V, Hara M, Hamlin A (1949) Effect of the lesion due to influenza virus on the resistance of mice to inhaled pneumococci. J Exp Med 89:53–68

Hastings JW, Nealson KH (1977) Bacterial bioluminescence. Annu Rev Microbiol 31:549–595

Hentzer M, Riedel K, Rasmussen TB, Heydorn A, Andersen JB, Parsek MR, Rice AS, Eberl L, Molin S, Givskov M (2002) Inhibition of quorum-sensing in *Pseudomonas aeruginosa* biofilm bacteria by a halogenated furanone compound. Microbiology 148:87–102

Heydorn A, Nielsen AT, Hentzer M, Sternberg C, Givskov M, Ersboll BK, Molin S (2000) Quantification of biofilm structures by the novel computer program COMSTAT. Microbiology 146:2395–2407

Hirsch P, Muller M (1986) Methods and sources for the enrichment and isolation of budding, nonprosthecate bacteria from freshwater. Microb Ecol 12:331–341

Hoiby N (2002) Understanding bacterial biofilms in patients with cystic fibrosis: current and innovative approaches to potential therapies. J Cyst Fibros 1:249–254

Hoiby N, Fomsgaard A, Jensen ET, Johansen HK, Kronborg G, Pedersen SS, Pressler T, Kharazmi A (1995) The immune response to bacterial biofilms. In: Lappin-Scott HM, Costerton JW (eds) Microbial Biofilms. Cambridge University Press, Cambridge, UK, pp 233–250

Hooper DU, Vitousek PM (1997) The effects of plant composition and diversity on ecosystem processes. Science 277:1302–1305

Hu J, Horn M, Limaye AP, Gautom RK, Fritsche TR (2002) Direct detection of Legionella in respiratory tract specimens by using Fluorescence In Situ Hybridization. In: Marre R (ed) Legionella. ASM Press, Washington, DC

Hurd H (2003) Manipulation of medically important insect vectors by their parasites. Annu Rev Entomol 48:141–161

Hurtle W, Shoemaker D, Henchal E, Norwood D (2002) Denaturing HPLC for identifying bacteria. Biotechniques 33:386–391

Jass J, Lappin-Scott HM (1996) The efficacy of antibiotics enhanced by electrical currents against *Pseudomonas aeruginosa* biofilms. J Antimicrob Chemother 38:987–1000

Jass J, Surman S, Walker J (2003) Medical biofilms, detection, prevention, and control. Wiley, New York

Jensen ET, Kharazmi A, Lam K, Costerton JW, Hoiby N (1990) Human polymorphonuclear leukocyte response to *Pseudomonas aeruginosa* grown in biofilms. Infect Immun 58:2383–2385

Kadouri D, O'Toole GA (2005) Susceptibility of biofilms to *Bdellovibrio bacteriovorus* attack. Appl Environ Microbiol 7:4044–4051

Kaiser D (1979) Social gliding is correlated with the presence of pili in *Myxococcus xanthus*. Proc Natl Acad Sci USA 76:5952–5956

Kaiser D (2004) Signaling in Myxobacteria. Annu Rev Microbiol 58:75–98

Khoury AE, Olson ME, Lam K, Nickel JC, Costerton JW (1989) Evaluation of the retrograde contamination guard in a bacteriologically challenged rabbit model. Br J Urol 63:384–388

Khoury AE, Lam K, Ellis B, Costerton JW (1992) Prevention and control of bacterial infections associated with medical devices. ASAIO J 38:M174–M178

Kim SK, Kaiser D, Kuspa A (1992) Control of cell density and pattern by intercellular signaling in Myxococcus development. Annu Rev Microbiol 46:117–139

Kjelleberg S (1993) Starvation in Bacteria. Plenum, New York

Klausen M, Aaes-Jorgensen A, Molin S, Tolker-Nielsen T (2003) Involvement of bacterial migration in the development of complex multicellular structures in *Pseudomonas aeruginosa* biofilms. Mol Microbiol 50:61–68

Koch R (1884) Die aetiologie der tuberkulose, mittheilungen aus dem kaiserlichen Gesundhdeitsamte 2:1–88

Kolenbrander PE, London J (1993) Adhere today, here tomorrow: oral bacterial adherence. J Bacteriol 175:3247–3252

Kolter R, Losick R (1998) All for one and one for all. Science 280:226–227

Korber DR, Lawrence JR, Lappin-Scott HM, Costerton JW (1995) A growth of microorganisms on surfaces. In: Lappin-Scott HM, Costerton JW (eds) Microbial Biofilms. Cambridge University Press, Cambridge, UK, pp 15–45

Kowalewska-Grochowska K, Richards R, Moysa GL, Lam K, Costerton JW, King EG (1991) Guidewire catheter change in central venous catheter biofilm formation in burn population. Chest 100:1090–1095

Krumbein WE, Paterson DM, Zavarzin GA (2003) Fossil and Recent Biofilms: A Natural History of Life on Earth. Springer, Berlin Heidelberg New York

Kudo H, Cheng K-J, Costerton JW (1987) Interactions between *Treponema bryantii* and cellulolytic bacteria in the in vitro degradation of straw cellulose. Can J Microbiol 33:244–248

Labbate M, Queck SY, Koh KS, Rice SA, Givskov M, Kjelleberg S (2004) Quorum sensing-controlled biofilm development in *Serratia liquefaciens* MG 1. J Bacteriol 186:692–698

Lam J, Chan R, Lam K, Costerton JW (1980) Production of mucoid microcolonies by *Pseudomonas aeruginosa* within infected lungs in cystic fibrosis. Infect Immun 28:546–556

Lambe DW Jr, Ferguson KP, Mayberry-Carson KJ, Tober-Meyer B, Costerton JW (1991) Foreign-body-associated experimental osteomyelitis induced with *Bacteroides fragilis* and *Staphylococcus epidermidis* in rabbits. Clin Ortho 266:285–294

Lamont RJ, Jenkinson HF (1998) Life below the gum line: pathogenic mechanisms of *Porphyromonas gingivalis*. Microbiol Mol Biol Rev 62:1244–1263

Lamont RJ, Chan A, Belton CM, Izutsu KT, Vasel D, Weinberg A (1995) *Porphyromonas gingivalis* invasion of gingival epithelial cells. Infect Immun 63:3878–3885

Lappin-Scott HM, Costerton JW (1995) Microbial Biofilms. Cambridge University Press, Cambridge, UK

Lawrence JR, Korber DR, Hoyle BD, Costerton JW, Caldwell DE (1991) Optical sectioning of microbial biofilms. J Bacteriol 173:6558–6567

Lawrence JR, Swerhone GDW, Leppard GG, Araki T, Zhang X, West MM, Hitchcock AP (2003) Scanning transmission X-Ray, laser scanning, and transmission electron microscopy mapping of the exopolymeric matrix of microbial biofilms. Appl Environ Microbiol 69:5543–5554

Lee W, Lewandowski Z, Nielsen PH, Hamilton WA (1995) Role of sulfate-reducing bacteria in corrosion of mild steel: a review. Biofouling 8:165–194

Leid JG, Shirtliff ME, Costerton JW, Stoodley P (2002) Human leukocytes adhere to, penetrate, and respond to *Staphylococcus aureus* biofilms. Infect Immun 70:6339–6345

Leung JWC, Ling TKW, Chan RCY, Cheung SE, Lai CW, Sung JJY, Chung SCS, Cheng AFB (1994) Antibiotics, biliary spesis, and bile duct stones. Gastrointest Endosc 40:716–721

Lewandowksi Z, Stoodley P (1995) Flow induced vibrations, drag force, and pressure drop in conduits covered with biofilm. Wat Sci Technol 32:19–26

Lewandowski Z, Altobelli SA, Fukushima E (1993) NMR and microelectrode studies of hydrodynamics and kinetics in biofilms. Biotechnol Prog 9:40–45

Lewandowski Z, Stoodley P, Altobelli S (1995) Experimental and conceptual studies on mass transport in biofilms. Wat Sci Technol 31:153–162

Lewis K (2001) The riddle of biofilm resistance. Antimicrob Agents Chemother 45:999–1007

Liu W, Smith DI, Rechtzigel KJ, Thibodeau SN, James CD (1998) Denaturing high performance liquid chromatography (DHPLC) used in the detection of germline and somatic mutations. Nucleic Acids Res 26:1396–1400

Long SR (2001) Genes and signals in the *Rhizobium*-Legume symbiosis. Plant Physiol 125:69–72

Majors PD, Minard KR, Ackerman EJ, Holtom GR, Hopkins DF, Weber TJ, Wind RA (2002) A combined confocal and magnetic resonance microscope for biological studies. Rev Sci Instrum 73:4329–4338

Maki DK, Weise CE, Sarafin HW (1977) A semi-quantitative method of identifying intravenous catheter-related infections. New Engl J Med 296:1305–1309

Marrie TJ, Costerton JW (1983) A scanning and transmission electron microscopy study of the surfaces of intrauterine contraceptive devices. Am J Obstet Gynecol 146:384–394

Marrie TJ, Costerton JW (1984) Scanning and transmission electron microscopy of in situ bacterial colonization of intravenous and intraarterial catheters. J Clin Microbiol 19:687–693

Marrie TJ, Nelligan J, Costerton JW (1982) A scanning and transmission electron microscopic study of an infected endocardial pacemaker lead. Circulation 66:1339–1341

Marsh PD, Bradshaw DJ (1995) Dental plaque as a biofilm. J Ind Microbiol 15:169–175

Marsh EJ, Luo H, Wang H (2003) A three-tiered approach to differentiate Listeria monocytogenes biofilm-forming abilities. FEMS Microbiol Lett 228:203–210

Marshall BJ, Warren JR (1984) Unidentified curved bacilli on the gastric epithelium of patients with gastritis and peptic ulceration. Lancet i:1311–1315

Marshall KC, Stout R, Mitchell R (1971) Mechanisms of the initial events in the sorbtion of marine bacteria to surfaces. J Gen Microbiol 68:337–348

Mashburn LM, Whiteley M (2005) Membrane vesicles traffic signals and facilitate group activities in a prokaryote. Nature 437:422–425

Mattick JS (2002) Type IV pili and twitching motility. Annu Rev Microbiol 56:289–314

May T, Ito A, Okabe S (2006) F-phenocopies: the characteristics of natural conjugative plasmid-bearing *Escherichia coli* biofilms. In: Proceedings of the 11th meeting of the International Society for Microbial Ecology. Abstract on page 173

Mayberry-Carson KJ, Tober-Meyer B, Smith JK, Lambe DW Jr, Costerton JW (1984) Bacterial adherence and glycocalyx formation in osteomyelitis experimentally induced with *Staphylococcus aureus*. Infect Immun 43:825–833

McGee ZA, Jensen RL, Clemens CM, Taylor-Robinson D, Johnson A, Gregg CR (1999) Gonococcal infection of human fallopian tube mucosa in organ culture: Relationship of mucosal tissue TNF-[alpha] concentration to sloughing of ciliated cells. Sexually Transmitted Dis 26:160–165

McLean RJ, Whiteley M, Stickler DJ, Fuqua WC (1997) Evidence of autoinducer activity in naturally occurring biofilms. FEMS Microbiol Lett 154:259–263

Mills J, Pulliam L, Dall L, Marzouk J, Wilson W, Costerton JW (1984) Exopolysaccharide production by viridans streptococci in experimental endocarditis. Infect Immun 43:359–367

Mitchell P, Moyle J (1965) Evidence discriminating between the chemical and the chemiosmotic mechanisms of electron transport phosphorylation. Nature 208:1205–1206

Møller S, Kristensen CS, Poulsen LK, Carstensen JM, Molin S (1995) Bacterial growth on surfaces: Automated image analysis for quantification of growth rate-related parameters. Appl Environ Microbiol 61:741–748

Møller S, Korber DR, Wolfaardt GM, Molin S, Caldwell FR (1997) Impact of nutrient composition on a degradative biofilm community. Appl Environ Microbiol 63:2432–2438

Morck DW, Raybould TJG, Acres SD, Babiuk LA, Nelligan J, Costerton JW (1987) Electron microscopic description of glycocalyx and fimbriae on the surface of *Pasturella haemolytica*. Can J Vet Res 51:83–88

Morck DW, Costerton JW, Bolingbroke DO, Ceri H, Boyd ND, Olson ME (1990) A guinea pig model of bovine pneumonic *Pasteurellosis*. Can J Vet Res 54:139–145

Moter A, Gobel UB (2000) Fluorescence in situ hybridization (FISH) for direct visualization of microorganisms. J Microbiol Methods 41:85–112

Moter A, Leist G, Rudolph R, Schrank K, Choi BK, Wagner M, Gobel UB (1998) Fluorescence in situ hybridization shows spatial distribution of as yet uncultured treponemes in biopsies from digital dermatitis lesions. Microbiology 144:2459–2467

Murga R, Foster TS, Brown E, Pruckler JM, Fields BS, Donlan RM (2001) Role of biofilms in the survival of *Legionella pneumophila* in model potable water system. Microbiology 147:3121–3126

Nealson KH (1997) Sediment bacteria: Who's there, what are they doing, and what's new? Annu Rev Earth Planetary Sci 25:403–434

Nealson KH, Saffarini D (1994) Iron and Manganese in anaerobic respiration: environmental significance, Physiology, and regulation. Annu Rev Microbiol 48:311–343

Nelson JW, Tredgett MW, Sheehan JK, Thornton DJ, Notman D, Govan JR (1990) Mucophilic and chemotactic properties of *Pseudomonas aeruginosa* in relation to pulmonary colonization in cystic fibrosis. Infect Immun 58:1489–1495

Nelson JL, Roeder BL, Carmen JC, Roloff F, Pitt WG (2002) Ultrasonically activated chemotherapeutic drug delivery in a rat model. Cancer Res 62:7280–7283

Newman DK, Banfield JF (2002) Geomicrobiology: how molecular scale interactions underpin biogeochemical systems. Science 296:1071–1077

Nguyen D, Singh PK (2006) Evolving stealth: genetic adaptation of *Pseudomonas aeruginosa* during cystic fibrosis infections. Proc Natl Acad Sci USA 103:8305–8306

Nichols WW (1991) Biofilms, antibiotics, and penetration. Rev Med Microbiol 2:177–181

Nickel JC, Ruseska I, Wright JB, Costerton JW (1985) Tobramycin resistance of cells of *Pseudomonas aeruginosa* growing as a biofilm on urinary catheter material. Antimicrob Agents Chemother 27:619–624

Nickel JC, Costerton JW, McLean JC, Olson M (1994) Bacterial biofilms: influence on the pathogenesis, diagnosis and treatment of urinary-tract infections. J Antimicrob Chemother 33:31–41

Nielsen PR, Lee WC, Lewandowski Z, Morrison M, Characklis W (1993) Corrosion of mild steel in an alternating oxic and anoxic biofilm system. Biofouling 7:267–284

Norris P, Noble M, Francolini I, Vinogradov AM, Stewart PS, Ratner BD, Costerton JW, Stoodley P (2005) Ultrasonically controlled release of ciprofloxacin from self-assembled coatings on poly(2-Hydroxyethyl Methacrylate) hydrogels for *Pseudomonas aeruginosa* biofilm prevention. Antimicrob Agents Chemother 49:4272–4279

Novitsky JA, Morita RY (1976) Morphological characterization of small cells resulting from nutrient starvation of a psychrophilic marine Vibrio. Appl Environ Microbiol 32:617–622

Olson ME, Lam K, Bosey GP, King EG, Costerton JW (1992) Evaluation of strategies for central venous catheter replacement. Crit Care Med 20:797–804

O'Toole GA, Kolter R (1998) Initiation of biofilm formation in *Pseudomonas fluorescens* WCS365 proceeds via multiple, convergent signaling pathways: a genetic analysis. Mol Microbiol 28:449–461

O'Toole GA, Kaplan HB, Kolter R (2000) Biofilm formation as microbial development. Annu Rev Microbiol 54:49–79

Oosthuizen MC, Steyn B, Theron J, Cosette P, Lindsay D, von Holy A, Brozel VS (2002) Proteomic analysis reveals differential protein expression by *Bacillus cereus* during biofilm formation. Appl Environ Microbiol 68:2770–2780

Palmer RJ, Gordon SM, Cisar JO, Kolenbrander PE (2003) Coaggregation-mediated interactions of *Streptococci* and *Acinomyces* detected in initial human dental plaque. J Bacteriol 185:3400–3409

Passador L, Cook JM, Gambello MJ, Rust L, Iglewski BH (1993) Expression of *Pseudomonas aeruginosa* virulence genes requires cell-to-cell communication. Science 260:1127–1130

Parsek MR, Singh PK (2003) Bacterial biofilms: an emerging link to disease pathogenesis. Annu Rev Microbiol 57:677–701

Parsek MR, Fuqua C (2004) Biofilms 2003: emerging themes and challenges in studies of surface-associated microbial life. J Bacteriol 86:4427–4440

Pelling AE, Li Y, Shi W, Gimzewski JK (2005) Nanoscale visualization and characterization of *Myxococcus xanthus* cells with atomic force microscopy. Proc Natl Acad Sci USA 102:6484–6489

Pesci EC, Milbank JB, Pearson JP, McKnight S, Kende AS, Greenberg EP, Iglewski BH (1999) Quinolone signaling in the cell-to-cell communication system of *Pseudomonas aeruginosa*. Proc Natl Acad Sci USA. 96:11229–11234

Post JC (2001) Direct evidence of bacterial biofilms in otitis media. Laryngoscope 111:2083–2094

Prigent-Combaret C, Vidal O, Dorel C, Lejeune P (1999) Abiotic surface sensing and biofilm-dependent regulation of gene expression in Escherichia coli. J Bacteriol 181:5993–6002

Purevdorj B, Costerton JW, Stoodley P (2002) Influence of hydrodynamics and cell signaling on the structure and behavior of *Pseudomonas aeruginosa* biofilms. Appl Environ Microbiol 68:4457–4464

Purevdorj-Gage L, Costerton JW, Stoodley P (2005) Phenotypic differentiation and seeding dispersal in non-mucoid and mucoid Pseudomonas aeruginosa biofilms. Microbiology 151:1569–1576

Raad I (1998) Intravascular-catheter-related infections. Lancet 351:893–898

Raghunathan A, Ferguson HR, Bornarth CJ, Song W, Driscoll M, Lasken RS (2005) Genomic DNA amplification from a single bacterium. Appl Environ Microbiol 71:3342–3347

Rayner MG, Zhang Y, Gorry MC, Chen Y, Post JC, Ehrlich GD (1998) Evidence of bacterial metabolic activity in culture-negative otitis media with effusion. J Am Med Assoc 279:296–299

Reed WP, Moody MR, Newman KA, Light PD, Costerton JW (1986) Bacterial colonization of hemasite access devices. Surgery 99:308–316

Rice AR, Hamilton MA, Camper AK (2003) movement, replication, and emigration rates of individual bacteria in a biofilm. Microb Ecol 45:163–172

Rivera ING, Chun J, Huq A, Sack RB, Colwell RR (2001) Genotypes associated with virulence in environmental isolates of *Vibrio cholerae*. Appl Environ Microbiol 67:2421–2429

Robinson DH (2005) Pleomorphic mammalian tumor-derived bacteria self-organize as multicellular mammalian eukaryotic-like organisms: morphogenetic properties in vitro, possible origins, and possible roles in mammalian tumor ecologies. Med. Hypotheses 63:177–188

Romero R, Espinoza J, Mazor M (2004) Can endometrial infection/inflammation explain implantation failure, spontaneous abortion, and preterm birth after in vitro fertilization? Fertil Steril 82:799–804

Rupp CJ, Fux CA, Stoodley P (2005) Viscoelasticity of *Staphylococcus aureus* biofilms in response to fluid shear allows resistance to detachment and facilitates rolling migration. Appl Environ Microbiol 71:2175–2178

Sadhu K, Domingue PAG, Chow AW, Nelligan J, Bartlett K, Costerton JW (1989) A morphological study of the in situ tissue-associated autochthonous microflora of the human vagina. Microb Ecol Health Dis 2:99–106

Sanderson SS, Stewart PS (1997) Evidence of bacterial adaptation to monochloramine in *Pseudomonas aeruginosa* biofilms and evaluation of biocide action model. Biotechnol Bioeng 56:201–209

Sanderson AR, Leid J, Hunsaker D (2006) Bacterial biofilms on the sinus mucosa of human subjects with chronic rhinosinusitis. Laryngoscope 116:1121–1126

Sauer K, Camper AK (2001) Characterization of phenotypic changes in *Pseudomonas putida* in response to surface-associated growth. J Bacteriol 183:6579–6589

Sauer K, Camper AK, Ehrlich GD, Costerton JW, Davies DG (2002) *Pseudomonas aeruginosa* displays multiple phenotypes during development as a biofilm. J Bacteriol 184:1140–1154

Schauder S, Shokat K, Surette MG, Bassler BL (2001) The LuxS family of bacterial autoinducers: biosynthesis of a novel quorum-sensing signal molecule. Mol Microbiol 41:463–476

Scheuerman TR, Camper AK, Hamilton MA (1998) Effects of substratum topography on bacterial adhesion. J Colloid Interface Sci 208:23–33

Schoolnik GK, Voskuil MI, Schnappinger D, Yildiz FH, Meibom K, Dolganov NA, Wilson MA, Chong KH (2001) Whole genome DNA microarray expression analysis of biofilm development by *Vibrio cholerae* O1 E1 Tor. Methods Enzymol 336:3–18

Selan L, Passariello L, Rizzo L, Varesi P, Speziale F, Renzini G, Thaller MC, Fioriani P, Rossolini GM (2002) Diagnosis of vascular graft infections with antibodies against staphylococcal slime antigens. Lancet 359:2166–2168

Shen KP, Antalis P, Gladitz J, Sayeed S, Ahmed A, Yu S, Hayes J, Johnson S, Dice B, Dopico R, Keefe R, Janto B, Chong W, Goodwin J, Wadowsky RM, Erdos G, Post JC, Ehrlich GD, Hu F (2005) Identification, distribution, and expression of novel genes in 10 clinical isolates of *Haemophilus influenzae*. Infect Immun 73:3479–3491

Shi W, Zusman DR (1993) The two motility systems of *Myxococcus xanthus* show different selective advantages on various surfaces. Proc Natl Acad Sci USA 90:3378–3382

Shirtliff ME, Leid JG, Costerton JW (2003) Basic science in musculoskeletal infections. In: Mader JT, Calhoun JH (eds) Musculoskeletal Infections. Dekker, New York, pp 1–61

Singh PK, Schaefer AL, Parsek MR, Moninger TO, Welch MJ, Greenberg EP (2000) Quorum-sensing signals indicate that cystic fibrosis lungs are infected with bacterial biofilms. Nature 407:762–764

Singh PK, Parsek MR, Greenberg EP, Welsh MJ (2002) A component of innate immunity prevents bacterial biofilm development. Nature 417:552–555

Sottile FD, Marrie TJ, Prough DS, Hobgood CD, Gower DJ, Webb LX, Costerton JW, Gristina AG (1986) Nosocomial pulmonary infection: possible etiologic significance of bacterial adhesion to endotracheal tubes. Crit Care Med 14:265–270

Spoering AL, Lewis K (2001) Biofilms and planktonic cells of *Pseudomonas aeruginosa* have similar resistance to killing by antimicrobials. J Bacteriol 183:6746–6751

Staley JT, Konopka A (1985) Measurement of in situ activities of nonphotosynthetic microorganisms in aquatic and terrestrial habitats. Annu Rev Microbiol 39:321–346

Stewart PS (1996) Theoretical aspects of antibiotic diffusion into microbial biofilms. Antimicrob Agents Chemother 40:2517–2522

Stewart PS, Costerton JW (2001) Antibiotic resistance of bacteria in biofilms. Lancet 358:135–138

Stoodley P, deBeer D, Lewandowski Z (1994) Liquid flow in biofilm systems. Appl Environ Microbiol 60:2711–2716

Stoodley P, deBeer D, Lappin-Scott HM (1997) Influence of electric fields and pH on biofilm structure as related to the bioelectric effect. Antimicrob Agents Chemother 41:1876–1879

Stoodley P, Lewandowski Z, Boyle JD, LappinScott HM (1998) Oscillation characteristics of biofilm streamers in turbulent flowing water as related to drag and pressure drop. Biotechnol Bioeng 57:536–544

Stoodley P, Dodds I, Lewandowski Z, Cunningham AB, Boyle JD, Lappin-Scott HM (1999a) Influence of hydrodynamics and nutrients on biofilm structure. J Appl Microbiol 85:19S–28S

Stoodley P, Boyle JD, deBeer D, Lappin-Scott HM (1999b) Evolving perspectives in biofilm structure. Biofouling 14:75–90

Stoodley P, Lewandowski Z, Boyle JD, Lappin-Scott HM (1999c) The formation of migratory ripples in a mixed species bacterial biofilm growing in turbulent flow. Environ Microbiol 1:447–457

Stoodley P, Wilson S, Hall-Stoodley L, Boyle JD, Lappin-Scott HM, Costerton JW (2001) Growth and detachment of cell clusters from mature mixed species biofilms. Appl Environ Microbiol 67:5608–5613

Stoodley P, Sauer K, Davies DG, Costerton JW (2002) Biofilms as complex differentiated communities. Annu Rev Microbiol 56:187–209

Suci PA, Mittelman MW, Yu FP, Geesey GG (1994) Investigation of ciprofloxacin penetration into Pseudomonas aeruginosa biofilms. Antimicrob Agents Chemother 38:2125–2133

Sullam PM, Drake TA, Sande MA (1985) Pathogenesis of endocarditis. Am J Med 78:110–115

Sung JY, Shaffer EA, Olson ME, Leung JW, Lam K, Costerton JW (1991) Bacterial invasion of the biliary system by way of the portal-venous system. Hepatology 14:313–317

Sung JY, Leung JWC, Shaffer EA, Lam K, Olson ME, Costerton JW (1992) Ascending infection of the biliary tract after surgical sphincterotomy and biliary stenting. J Gastroenterol Hepatol 7:240–245

Sung JY, Leung JWC, Shaffer EA, Lam K, Costerton JW (1993) Bacterial biofillm, brown pigment stone and blockage of biliary stents. J Gastroenterol Hepatol 8:28–34

Sutherland IW (1977) Surface Carbohydrates of the Prokaryotic Cell. Academic, London

Szomolay B, Klapper I, Dockery J, Stewart PS (2005) Adaptive responses to antimicrobial agents in biofilms. Environ Microbiol 7:1186–1191

Tenney JH, Moody MR, Newman KA, Schimpff SC, Wade JC, Costerton JW, Reed WP (1986) Adherent microorganisms on lumenal surfaces of long-term intravenous catheters: importance of Staphylococcus epidermidis in patients with cancer. Arch Internal Med 146:1949–1954

Thar R, Kühl M (2002) Conspicuous veils formed by vibroid bacteria in sulfidic mine sediment. Appl Environ Microbiol 68:6310–6320

Tolker-Nielsen T (2006) Physiological differentiation in Pseudomonas aeruginosa biofilms. Proc Int Soc Microb Ecol, Abstract 163

Tolker-Nielsen T, Brinch UC, Ragas PC, Andersen JB, Jacobsen CS, Molin S (2000) Development and dynamics of Pseudomonas sp. biofilms. J Bacteriol 182:6482–6489

van Loosdrecht MC, Norde W, Zehnder AJ (1990) Physical chemical description of bacterial adhesion. J Biomater Appl 5:91–106

Veeh RH, Shirtliff ME, Petik JR, Flood JA, Davis CC, Seymour JL, Hansmann MA, Kerr KM, Pasmore ME, Costerton JW (2003) Detection of Staphylococcus aureus biofilm on tampons and menses components. J Infect Dis 188:519–530

Vorachit M, Lam K, Jayanetra P, Costerton JW (1995a) The study of the pathogenicity of Burkholderia pseudomallei – a guinea pig model. J Infect Dis Antimicrob Agents 12:115–121

Vorachit M, Lam K, Jayanetra P, Costerton JW (1995b) Electron microscopy study of the mode of growth of Pseudomonas pseudomallei in vitro and in vivo. J Trop Med Hyg 98:379–391

Wagner M, Horn M, Daims H (2003) Fluorescence in situ hybridization for the identification and characterization of prokaryotes. Curr Opin Microbiol 6:302–309

Wall D, Kaiser D (1999) Type IV pili and cell motility. Mol Microbiol 32:1–10

Ward KH, Olson ME, Lam K, Costerton JW (1992) Mechanism of persistent infection associated with peritoneal implants. J Med Microbiol 36:406–413

Ward DM, Ferris MJ, Nold SC, Bateson MM (1998) A natural view of microbial biodiversity within hot spring cyanobacterial mat communities. Microbiol Mol Biol Rev 62:1353–1370

Webb JS, Thompson LS, James S, Charlton T, Tolker-Nielsen T, Koch B, Givskov M, Kjelleberg S (2003) Cell death in Pseudomonas aeruginosa biofilm development. J Bacteriol 185:4585–4592

Wellman N, Fortun SM, McLeod BR (1996) Bacterial biofilms and the bioelectric effect. Antimicrob Agents Chemother 40:2012–2014

Whitchurch CB, Tolker-Nielsen T, Ragas PC, Mattick JS (2002) Extracellular DNA required for bacterial biofilm formation. Science 295:1487

Woods DE, Straus DC, Johanson WG Jr, Bass JA (1981) Role of fibronectin in the prevention of the adherence of Pseudomonas aeruginosa to buccal cells. J Infect Dis 143:784–790

Wu H, Song Z, Hentzer M, Andersen JB, Heydorn A, Mathee K, Moser C, Eberl L, Molin S, Hoiby N, Givskov M (2000) Detection of N-acylhomoserine lactones in lung tissues of mice infected with Pseudomonas aeruginosa. Microbiology 146:2481–2493

Wu H, Song Z, Hentzer M, Andersen JB, Molin S, Givskov M, Hoiby N (2004) Synthetic furanones inhibit quorum sensing and enhance bacterial clearance in Pseudomonas aeruginosa infections in mice. J Antimicrob Chemother 53:1054–1061

Wullt B, Connell H, Rollano P, Mansson W, Coleen S, Svanborg C (1998) Urodynamic factors influence the duration of Escherichia coli bacteriuria in deliberately colonized cases. J Urol 159:2057–2062

Wyndham RC, Costerton JW (1981) In vitro microbial degradation of bituminous hydrocarbons and in situ colonization of bitumen surfaces within the Athabasca oil sands deposit. Appl Environ Microbiol 41:791–800

Wyndham RC, Cashore AE, Nakatsu CH, Peel MC (1994) Catabolic transposons. Biodegradation 5:323–342

Xavier KB, Bassler BL (2003) LuxS quorum sensing: more than just a numbers game. Curr Opin Microbiol 6:191–197

Xie H, Cook GS, Costerton JW, Bruce G, Rose TM, Lamont RJ (2000) Intergeneric communication in dental plaque biofilms. J Bacteriol 182:7067–7069

Zhong W, Millsap K, Bialkowska-Hobrzanska H, Reid G (1998) Differentiation of Lactobacillus species by molecular typing. Appl Environ Microbiol 64:2418–2423

ZoBell CE (1943) The effect of solid surfaces upon bacterial activity. J Bacteriol 46:39–56

Suggested Reading

Caldwell DE, Costerton JW (1996) Are bacterial biofilms constrained to Darwin's concept of evolution through natural selection? Microbiologica 12:347–358

Costerton JW (2005) Biofilm theory can guide the treatment of device-related orthopedic infections. Clin Orthop Rel Res 437:7–11

Costerton JW, Stewart PS (2001) Battling biofilms. Sci Am 285:75–78

Costerton JW, Geesey GG, Cheng GK (1978) How bacteria stick. Sci Am 238:86–95

Costerton JW, Cheng K-J, Geesey GG, Ladd TI, Nickel JC, Dasgupta M, Marrie TJ (1987) Bacterial biofilms in nature and disease. Annu Rev Microbiol 41:435–464

Costerton JW, Lewandowski Z, Caldwell DE, Korber DR, Lappin-Scott HM (1995) Microbial biofilms. Annu Rev Microbiol 49:711–745

Costerton JW, Stewart PS, Greenberg EP (1999) Bacterial biofilms: a common cause of persistent infections. Science 284:1318–1322

Davies D (2003) Understanding biofilm resistance to antibacterial agents. Nat Rev Drug Discov 2:114–122

Donlan RM (2001) Biofilm formation: a clinically relevant microbiological process. Clin Infect Dis 33:1387–1392

Donlan RM, Costerton JW (2002) Biofilms: survival mechanisms of clinically relevant microorganisms. Clin Microbiol Rev 15(2):167–193

Douglas LJ (2003) Candida biofilms and their role in infection. Trends Microbiol 11:30–36

Drenkard E, Ausubel FM (2002) *Pseudomonas* biofilm formation and antibiotic resistance are linked to phenotypic variation. Nature 416:740–743

Filloux A, Vallet I (2003) Biofilm: set-up and organization of a bacterial community. Med Sci Paris 19:77–83 (in French)

Fuqua WC, Greenberg EP (2002) Listening in on bacteria: acyl-homoserine lactone signaling. Nat Rev Mol Cell Biol 3:685–695

Fux CA, Stoodley P, Hall-Stoodley L, Costerton JW (2003) Bacterial biofilms: a diagnostic and therapeutic challenge. Expert Rev Anti-Infect Ther 1:667–683

Fux CA, Costerton JW, Stewart PS, Stoodley P (2005) Survival strategies of infectious biofilms. Trends Microbiol 13:34–40

Ghannoum M, O'Toole GA (2004) Microbial biofilms. ASM, Washington, DC, pp 1–426

Gibbons RJ, van Houte J (1975) Dental caries. Annu Rev Med 26:121–136

Gilbert P, Maira-Litran T, McBain AJ, Rickard AH, Whyte FW (2002) The physiology and collective recalcitrance of microbial biofilm communities. Adv Microbiol Physiol 46:202–256

Hall-Stoodley L, Costerton JW, Stoodley P (2004) Bacterial biofilms: from the natural environment to infectious diseases. Nat Rev Microbiol 2:95–108

Hoiby N (2002) Understanding bacterial biofilms in patients with cystic fibrosis: current and innovative approaches to potential therapies. J Cyst Fibros 1:249–254

Jass J, Surman S, Walker J (2003) Medical biofilms, detection, prevention, and control. Wiley, New York

Kaiser D (2004) Signaling in Myxobacteria. Annu Rev Microbiol 58:75–98

Kariyama R, Kumon H (2003) Biofilm infections. Nippon Rinsho 61:266–271 (in Japanese)

Kjelleberg S (1993) Starvation in Bacteria. Plenum, New York

Kjelleberg S, Molin S (2002) Is there a role for quorum sensing signals in bacterial biofilms? Curr Opin Microbiol 5:254–258

Kolter R, Losick R (1998) All for one and one for all. Science 280:226–227

Krumbein WE, Paterson DM, Zavarzin GA (2003) Fossil and Recent Biofilms: A Natural History of Life on Earth. Springer, Berlin Heidelberg New York

Lamont RJ, Jenkinson HF (1998) Life below the gum line: pathogenic mechanisms of *Porphyromonas gingivalis*. Microbiol Mol Biol Rev 62:1244–1263

Lappin-Scott HM, Costerton JW (1995) Microbial Biofilms. Cambridge University Press, Cambridge, UK

Lindow SE, Hecht-Poinar EI, Elliot VJ (2002) Phyllosphere Microbiology. APS, St Paul, MN

Mateo MM, Maestre VJR (2004) Biofilm: model of bacterial communication and resistance to antimicrobial agents. Rev Esp Quimioter 17:26–28 (in Spanish)

Morris DP, Hagr A (2005) Biofilm: Why the sudden interest? J Otolaryngol Suppl 2:S56–S59

O'Toole GA, Kaplan HB, Kolter R (2000) Biofilm formation as microbial development. Annu Rev Microbiol 54:49–79

Palmer RJ (2004) Peter Hirsch and biofilms: microbial ecology's role in a new field. Microb Ecol 47:200–204

Parsek MR, Fuqua C (2004) Biofilms 2003: emerging themes and challenges in studies of surface-associated microbial life. J Bacteriol 86:4427–4440

Parsek MR, Singh PK (2003) Bacterial biofilms: an emerging link to disease pathogenesis. Annu Rev Microbiol 57:677–701

Potera C (1996) Biofilms invade microbiology. Science 273:1795–1797

Raad I (1998) Intravascular-catheter-related infections. Lancet 351:893–898

Ramadan HH, Sanclement JA, Thomas JG (2005) Chronic rhinosinusitis and biofilms. Otolaryngol Head Neck Surg 132:414–417

Ramey BE, Koutsoudis M, von Bodman SB, Fuqua C (2004) Biofilm formation in plant-microbe associations. Curr Opin Microbiol 7:602–609

Slavkin HC (1997) Biofilms, microbial ecology, and Antonie van Leeuwenhoek. J Am Dent Assoc 128:492–495

Stewart PS, Costerton JW (2001) Antibiotic resistance of bacteria in biofilms. Lancet 358:135–138

Stoodley P, Sauer K, Davies DG, Costerton JW (2002) Biofilms as complex differentiated communities. Annu Rev Microbiol 56:187–209

Sutherland IW (1977) Surface Carbohydrates of the Prokaryotic Cell. Academic, London

Webb JS, Givskov M, Kjelleberg S (2003) Bacterial biofilms: prokaryotic adventures in multicellularity. Curr Opin Microbiol 6:578–585

Wilson M (2005) Microbial Inhabitants of Humans: Their Ecology and Role in Health and Disease. Cambridge University Press, Cambridge, UK

Wilson M, Devine D (2003) Medical Implications of Biofilms. Cambridge University Press, Cambridge, UK, pp 1–173

Subject Index

Printing: Krips bv, Meppel
Binding: Stürtz, Würzburg